AT CROSSROADS WITH CHICKENS

AT
CROSSROADS
with Chickens

A "What If It Works?" Adventure
in Off-Grid Living & Quest for Home

Tory McCagg

With thanks –
Tory

BAUHAN PUBLISHING ✦ PETERBOROUGH ✦ NEW HAMPSHIRE

2020

Library of Congress Cataloging-in-Publication Data

Names: McCagg, Tory, author.
Title: At crossroads with chickens : a "what if it works?" adventure in off-grid living & quest for home / Tory McCagg.
Description: Peterborough, New Hampshire : Bauhan Publishing, 2020. |
Summary: "In 2012, McCagg and her husband built a solar-powered house in New Hampshire. It was to be a weekend getaway, with cats and chicks in tow, but they ended up moving there permanently when they learned their rooster would be banned back home in Rhode Island. While chicken kerfuffles lighten the mood, this is a story born of heartbreak, of yearning for the great beauty of the world as it used to be. McCagg interlaces her tale with her mother's battle with Parkinson's, braiding both Mother and Mother Nature in this perfect storm of personal growth rippling out to effect a larger transformation"— Provided by publisher.
Identifiers: LCCN 2020007380 (print) | LCCN 2020007381 (ebook) |
ISBN 9780872333178 (trade paperback) | ISBN 9780872333185 (ebook)
Subjects: LCSH: McCagg, Tory. | Farm life—New Hampshire. | Solar houses—New Hampshire—Biography.
Classification: LCC S521.5.N4 M33 2020 (print) | LCC S521.5.N4 (ebook) | DDC 630.9742--dc23
LC record available at https://lccn.loc.gov/2020007380
LC ebook record available at https://lccn.loc.gov/2020007381

Book Design by Henry James, set in Bembo Book Pro with Fairbank titles
Cover Design by Henry James
Cover photograph and all interior photographs by the author
Front cover chicken by Hartwig Kopp-Delaney, used by permission

Printed by Versa Press

Tory can be reached at www.torymccagg.com

BAUHAN PUBLISHING LLC
PO BOX 117 PETERBOROUGH NEW HAMPSHIRE 03458
WWW.BAUHANPUBLISHING.COM
603-567-4430

Follow us on Facebook and Twitter – @bauhanpub

Printed in the United Sates of America

To my mother and, of course, Carl

If I can stop one heart from breaking, I shall not live in vain.
If I can ease one life the aching, Or cool one pain,
Or help one fainting robin Unto his nest again,
I shall not live in vain.

EMILY DICKINSON

In order to attain the impossible, one must attempt the absurd.

MIGUEL DE CERVANTES

We don't have to heal the earth; she can heal herself.
All we have to do is stop making her sick.

WALLACE BLACK ELK, LAKOTA SHAMAN

CONTENTS

Rainbow between us and Mt. Monadnock

I

THE WEB WE WOVE

Whereas Mount Monadnock has an unimpeded view
of all six New England states, my view has impediments:
trees, other mountains, my assumptions and expectations.

ROOTS

In 1966–67, when I was four, my family lived in Hungary for a year. During that time my father, an East European historian, researched his book *Jewish Nobles and Geniuses*. While he sequestered himself in the Jewish archives, my artist mother painted, met other artists, and took care of my sister, Xanda, and me. My parents' social life revolved around an *avant-garde*, intellectual circle that was watched by the Communist regime. The only conscious memory of that time that I have is getting onto a crowded bus with Xanda and a babushka.

Or maybe, like most of my childhood memories, that's just a photograph. Other photos of that time include my parents sitting with their dear friends, Miklós and Szuszi Erdély, in their kitchen, smoking cigarettes and discussing the artistic and political happenings of the day; the Erdélys' teenaged sons, Dani and Yuri, standing outside the gate to the enclosed garden of their decaying home; and my sister and me playing with a pile of Dachshund puppies. I wonder: at such a young age, did I take in that country's fear, its history of repression and terror, a sense of being watched, overseen, and judged?

But wait; those photos might be of a later time, return trips. And does it matter how old I was—four? seven?—when my parents took me to the countryside, or even if we were in Hungary? In my mind's eye, my parents and I stand under a grape arbor at a country villa surrounded by switchback paths and fields. The memory doesn't include my sister. Two-and-a-half years older, Xanda had perhaps stayed with friends. There was a girl there though, slightly older than I. I see her in a light cotton dress, long-limbed and slender, with black braids past her shoulders. She was going to walk to a lake for a swim and to pick beautiful, blue flowers. Did I want to come?

I was not entirely convinced, uneasy, perhaps, in the strangeness of the country, the foreignness of the language. My parents, though, encouraged me, likely as not baffled by my limited curiosity. I went. The girl and I walked along those switchback paths, on and on into the unfamiliar. It seemed a terribly long way to that beautiful lake.

"We'll be there soon," the girl promised me in stilted English, over and over again. I squinted to see in the blinding sunlight. My clothes stuck to my skin. The heat suffocated any sounds as every step increased the distance between me and safety, my parents, and home. I was very far from our home in East Lansing, Michigan. Why was I not there, curled on a couch and reading *The Story of Dr. Dolittle* to my cats, Foggy and Bathsheba? Instead, I followed a scorching path with trampled, yellowed grass behind a strange girl, who walked and walked and promised a lake of cool water. Where was it? Did it exist? I got scared, remembering the stories of gypsies kidnapping children, breaking their legs and leaving them to beg in the streets. Maybe this girl, taking me so far away, was a gypsy. Were we really going to a lake? I doubted it. I wanted to go back. The girl looked at me. "I'm going to the lake to swim and pick blue flowers."

She continued on. I turned back. The road looked different. The switchback lanes rising up to my right and down to my left shimmered with heat. I hesitated . . . but the girl was gone, and so I walked alone. I was lost and weeping when an old peasant (who to my eye looked remarkably like a gypsy with his grizzled beard and dirt-covered clothes) came up with his pony and cart. He spoke no English. But even if he had, I wouldn't have known what to say, where I was, or where I needed to go. Somehow he knew. Maybe he had seen me with the girl and knew where she lived. He took me back to my parents. They were surprised to see me back so soon and did not understand my tears.

Flash memories. Influences of the past. Why begin with this anecdote, full of doubtful facts? Because it shows these two things: at that young age, absence filled my heart, and it took a farmer to save me.

Driveway in summer

THE PLACE AND THE PLAYERS

Darwin's View, like Mount Monadnock herself, is at the meeting point of Dublin and Jaffrey, New Hampshire. Though both towns lay claim to her, the mountain's home is in Jaffrey. So is mine.

It has taken me a long time to get to the point that I can accept that: Darwin's View is my home. What began as a 24-by-38-foot timber-frame house can be found at the end of what has been variously called a logging road, Catch-22 Drive, Mud Mountain, and Heartbreak Hill. If you follow the rise of that half-mile sinuous lane, cross over a barely discernible wetland, continue up and around an entirely legal, if remarkably steep, 9.2-grade curve, you will pass by my husband, Carl's, budding orchard and hops-festooned flag pole and arrive at the top. There you will experience our panoramic view: from north of North Pack Monadnock to Mount Watatic, westward along the hills of the New Hampshire–Massachusetts border, ending with the entirety of Mount Monadnock. Breathtaking.

Especially if the wind is blowing, as it so often does, over the mountain, down into the valley, and up, smack into us.

The wind, as representative of Mother Nature at her most willful and unforgiving, is a force to be reckoned with here, if only because Carl and I tend to remember too late how far unanchored items will fly. On blowy days, we have been found chasing a variety of objects across the field: a portable hot tub that had been waiting to be repurposed; Styrofoam insulated panels, ditto; two chicken runs; a brand-new pair of prescription eyeglasses; hay bales. Thus far, not the chickens. Though they are buffeted, too, by the wind, the chickens—as opposed to the humans—make sure to keep themselves snugly in their coop when the wind nears gale proportions. Even so, I have lain awake listening to it howl, feeling the house shudder, convinced that this is my Dorothy of Oz moment and the chickens are about to play the part of Toto. More than once, I have had to remind myself that this is not Kansas. Tornadoes don't happen in the mountains.

Or do they?

Tornadoes are formed by rising warm air and dropping cold air twisting together into spirals. They are funnels of focused chaos that destroy anything and everything in their path. Flat plains to skyscrapers and mountains, nothing stops a tornado's hunger for destruction. Its whims are Mother Nature's, and she, as I have learned, does not care. She throws her fury, lashes out her outrage like a cat-o'-nine-tails whip. The consequence of the violence is all equal to her because she is a woman scorned. She is outraged. There is no safe place, no hope.

Full stop. If Carl had his way, he would stop me right about now.

It happens on a daily basis. I get upset with myself, or the world around me, and go on a bender of—dare I say it?—negativity. In reaction to yet another outrage in the news, or a passing train of thought that I hitch onto, I might go so far as to reject an open-faced avocado sandwich that Carl has lovingly prepared for me. He, in turn, will study me with his New Hampshire blue eyes, shake his head, and admonish me.

"Don't blame your lunch, Tory. It's not its fault."

I berate myself because he is right. That's like blaming cows for the world's methane problem.

Nicknames Carl has earned over the years: "Iron Man," because he is a mind-over-body kind of guy; "White Whale," with his burly body and burn-in-the-sun skin; "Caliachee," toned with the fondness and love that those who know him feel for this playful fellow who is ever willing to help, fix, care. More recently, he has earned the moniker of "Compost Carl." Besides his trombone and me, his passion is returning what came from the earth to the earth. Carl, with his long, blond-to-white eyebrows and dimples that promise jovial good fellowship. All this as opposed to me.

I called a friend about exactly this contrast. I was in the midst of a fit of anxiety, having endured yet another social situation during which Carl, somehow, found topics of conversation to discuss while I stood mum and smiling stiffly, stymied by what to say, thereby leaving the impression of cool aloofness, possibly unfriendliness or, horrors, rudeness.

"Am I at all likable?" I asked Deej, who has known me since we were both in our early twenties. He laughed, kindly, and not because he didn't realize how close to tears I was but because he did know that, and found it endearing.

"Carl's like an open book," Deej said. "He's accessible, and full of ideas. You're more like a wetsuit. A person has to pull you on, test you out, see how you fit. It's more of a commitment, getting to know you."

I had never been called a wetsuit before. But it made perfect sense. In fact, I liked the idea because it worked both ways. People had to try me out but that extra skin protected me, kept people away until I could trust them.

In the meantime, my wetsuit and I *appear* to be confident, organized, and directed. My lithe (if very tense) body, rosy (if gaunt-cheeked) complexion, and seeming ability to create a meal out of whole cloth and a home where no home should be cause people to expect Martha Stewart, not . . . me. As one childhood friend of Carl's recently said, "When I first met you, I thought you had your shit together. Then I got to know you."

Yes. When they get to know me, they find that I am disorder personified and, according to many of his childhood friends, the only one nice enough to deserve Carl. I pass, a helium balloon tethered to earth by a string that Carl holds. He grounds me, even as I distract.

Except when he distracts. Then I ground him.

Thus, together, we resemble a tornado. Two forces of nature, we have wrought destruction with an eye toward rebuilding and healing. Three decades into our marriage, we've only just begun.

17

Hopeful? That is to be determined. But even Carl can't change this fact: my real name is Dorothy. Like Dorothy Gale, I believe there is no place like home. But isn't home supposed to feel safe and secure, stable? In which case, what am I doing on this windswept hill, transplanted at the age of fifty to unfamiliar environs and unwonted pastimes?

My peripatetic childhood took me to Europe, but mostly we traveled between East Lansing, Michigan, where my father taught, and Connecticut, where both my parents' families had settled, Hartford and Stonington, respectively. It was in Stonington that I learned the ocean's language. I walked on beaches, sand crunching under my feet as I picked perfect seashells and admired the prehistoric carapace of dead horseshoe crabs. Wading into the cool shallows of the harbor's water, I would stand still as minnows darted past my ankles, tickling my toes. Seagrass caught and clung to my calves. The damp salt air frizzed my hair, coating my skin with salt and sand and filling me with the smells of fish and brine and seaweed. The call of a foghorn. The cries of seagulls, terns, and cormorants. Tumultuous waves balancing calm waters, like yin and yang encircling, splashing on the beach. The ocean, with its ebb and flow, is about motion. Into oceans, one can immerse oneself, float, swim, or drown.

Mount Monadnock speaks a different language. Some days, it is lost to me in clouds or the blindness of routine. Other days, I look to Mount Monadnock every few minutes. Alternately veiled by shades of blue and purple haze, or exposed by the brilliance of the sun, the subtlety of the moon, Monadnock tries on a different mood with each passing moment. She is solidly there. Mountains

Me, at about age one

18

don't move; rather, they move only on glacial time scales. Thus, mountains counter an ocean's movement with stability and permanence. Mountains express the resonance of soil and stone.

Majestic. Beautiful. These are overused words that don't evoke Mount Monadnock's essence, her timeless magnetism that attracts artists and writers, musicians and dancers, hikers, leaf peepers, long-time residents. So many return, again and again, to be in her shadow or ascend her rocky shoulders. Why?

Some come to follow in the footsteps of the famous: transcendentalists Ralph Waldo Emerson, Henry David Thoreau, and Margaret Fuller. Painters John Singer Sargent and Abbott Handerson Thayer. Writers Mark Twain, Thornton Wilder, and Willa Cather. Cather wrote some of her most famous works here, and is buried in Jaffrey Center with her long-time "companion" Edith Lewis. And the musicians: Leonard Bernstein, Aaron Copeland, Edward and Marian MacDowell. These last two established the MacDowell Colony, which boasts of having supported 7,700 artists. The Monadnock region has been, and is, alive with the creative ilk.

Other people come, perhaps, because they have heard the mountain is challenging to scale but accessible. They can climb her and return down in an afternoon, and only occasionally does the effort result in a broken leg or heart attack. Some people climb her every day. Some once a year. Some to get engaged, and others to spread a loved one's ashes. There have been clubs formed for annual walks up. Until 1954, when it burned down, people stayed at the rough-and-tumble hotel the Halfway House. And in Carl's youth there was a hut at the top that he remembers, vaguely, as a concession stand. People, and more people, come here. Is it to be in the woods, experiencing and learning from nature, or for the stupendous panorama that, on a clear day, extends to the John Hancock building in Boston and the Green Mountains of Vermont?

From our distance at Darwin's View, Carl and I note Mount Monadnock's presence as she is profiled by the sunrise's reflective blush or the evanescence of a fiery sunset. Terrible and joyous, she penetrates our hearts. She dares Carl, if not me, to climb her, because the mountain's syntax is best communicated when climbing, grasping the stone, and pulling up, up to see the marriage of granite and flora as trees sit on rocks, embracing.

I will note here that, because so many climb her, there is the suggestion that she is easy to summit. This is a baldfaced lie. At 3,165 feet above sea level, and with 300 feet of that rise being above the timberline, there is a lot of granite to scale. Granite is hard. It is rock and Mount Monadnock—be she muse, classroom, or maker of great wind—is called a mountain for a reason.

A beloved mountain. I would suggest that almost every one of the one hundred thousand–plus people who come here each year have her hook in their hearts when they leave. She calls to them as a symbol of the spiritual and natural beauty of this region. And for those who have grown up here and returned, she holds a special place.

Carl's Huck Finnian youth in Peterborough, New Hampshire, the town immediately northeast of Jaffrey, was full of escapades on and around Mount Monadnock. He and his friends climbed the mountain, played on her shoulders, at her feet. I have heard stories of their escapades, time and again, and am happy to hear them repeated because those who join in the reminiscing had such fun, such joy. The mountain is a part of them, of Carl. Each morning, as he watches the sunrise turning the mountain

Carl as a kid, with Mount Monadnock behind him

pink, his heart responds. He breathes to her rhythms. Until we moved to Jaffrey, I didn't understand that. Carl's heart was here. And climbing the mountain, like skiing, is an event that Carl anticipates and adores.

I, on the other hand, tend toward vertigo and hyperventilation at the very idea of either sport. Skiing and mountain climbing are Exhibit A *dangerous*. I will walk quite happily on a paved surface, but put me at the top of a mountain on skis or at the bottom of one in hiking boots, and I am out of my comfort zone. Thus, I was less than enthusiastic when I first attempted to climb Mount Monadnock with Carl, early on in our marriage. I was recovering from bronchitis. Or that was my excuse when we reached the false peak and I gave up. It was a mind-over-body moment and my body, for once, won. Too steep. Too endless. We had only been an hour or two but I was in my "there are too many trees around me, get me out of here" phase. In short, I had my limits, which were clearly more limiting than those of all the tiny tots hopping and skipping past us, along with their huffing, beet-faced parents.

Years later, in 2009, we rented an apartment in Dublin, New Hampshire, the town where Carl's sister and brother-in-law live. There Mary and Tom had raised their two children, Sarah and Jared, who in turn were raising their children. Dublin is the town north of Jaffrey, and Jaffrey was where we had bought 193 acres of land three years earlier with vague plans, that had yet to come into fruition, of building a place off-grid. In the meantime, perhaps a yurt.

Unfortunately, as we came to find out, the town of Jaffrey wouldn't allow us to put up a yurt, also known as a tent, because in Jaffrey, a yurt had to be connected to a septic system, thereby defeating the purpose of a tent. Too bad. We had wanted to get a sense of the land before putting in anything permanent—like, say, a dome space house that rotates when you press its remote control. We would be able to turn the entire house to face the sun, or away, or toward any of our varying views. It's the coolest of passive solar designs, unless you lose the remote. That being a high-risk possibility, us being us, we moved on to an even better idea: a hobbit house. In the ground. Organic. *Wind*-proof with its sod roof. No one would know we were there; it would be so much a part of the landscape. But the negative Nellies around us warned that we would be in for trouble if the roof leaked. And never forget: eventually all roofs leak.

Whatever house we built, we faced this formidable fact: we had been married for nearly twenty years and I had never reached the peak of Mount Monadnock. We hadn't had the time. But in that Dublin, New Hampshire, apartment, we were living at the bottom of it. How could we not go to the top? Up we got and off we went, with me feeling as I always do when I step out of my routine: full of dread and distinctly uncomfortable with the task before me. I berated myself for such cowardice. I claimed to like exercise and physical challenge. Here was my opportunity. But unlike running—in my teens and twenties I would run for miles—this unfamiliar physical challenge made me nervous. What if I fell off a cliff? Got lost? Failed to reach the summit again?

Our apartment wasn't even a quarter of a mile from the bottom of the Pumpelly Trail, at four-and-a-half miles the longest route to the top of Mount Monadnock. As we headed out, our knapsacks packed with water, snacks, and sandwiches, I knew it was an in-for-a-penny, in-for-a-pound kind of day. We entered the woods and began walking along the cool, forested path. Pine needles softened our footfalls. Tree roots on occasion tripped us. A gradual rise and soon enough we were climbing, and scaling that trail, and it is truly remarkable to me the variety of people (and dogs) who manage to ascend the mountain and survive. How did women wearing corsets and straw hats bedecked with floral arrangements do it? I wondered as my long-dormant calf and thigh muscles called out for attention.

I stepped on and up the devil's ladder. Mossy boulders made way for circuitous rocky paths. We paused at an opening of the woods to admire the view. Or, rather, Carl looked and located his past. In one direction, he pointed to the general vicinity of his grandmother's old house; in another, to Lake Skatutakee—what a fun word to say—and, over in the distant hinterlands, to some mountain he had climbed or skied. I don't know which, because I had turned my back to the view to hydrate. The approximately 65 percent of me that is water had exited my body, soaking my T-shirt, leaving the inside of me parched and the outside of me salty and slightly chilled. We were very high up, and proceeded to go higher. The end of the hike required clambering over boulders and crevices, with only scrawny shrubs to cling to—and do not forget not to look down if you tend toward vertigo.

Looking up through sweat and, at times, breathless terror, I made it to the top. I felt triumphant, if a bit bruised, and was not alone in that: scores of other summiteers shared with us the unobstructed, 360-degree view. And because the top of Mount Monadnock is relatively wide and without too much sense of sheer cliff, a feeling of cavalier safety set in. I could admire the landscape, the clouds in the sky, the day. Hawks glided and swooped on the passing breezes. The children shrieking and running about, heading straight for one or another precipice, were not my problem. Instead, I channeled the (prohibited, but there they were) dogs flopped ecstatically on the warm stone, panting with joy. Carl and I, too, flopped down on the lichen-covered granite to rest and eat our sandwich lunch. We tried to determine where our 193 acres of land might be. Binoculars would have come in handy. There were various fields and a pond or three to choose from. An iota of excitement shimmered in me. Somewhere out there was the place we would create: Darwin's View.

We had named the property Darwin's View when we put the majority of it into a conservation easement the year we bought it, 2006. That first year had been a rough-and-tumble effort to get various permits and easements. Darwin's View, with its suggestion of evolution, effort over the long haul, and survival, seemed apt. This would not be an easy place. It would be bigger than us and here for longer. At least, that's what we hoped. We wanted to make something of it that would make a difference.

Inspired by the future, we ate our picnic lunch and returned by way of the Dublin Trail. The Farmer's Trail, as it is sometimes called, is short and steep: 1.9 miles. We headed down, straight down; the trail and I, apparently, preferred going up. Certainly, my knees did. My knees thought they had been given a well-deserved retirement when I gave up running, and were quite vocal in their protest as we bashed and pounded down the trail.

Pain, I learned, can be like electricity, the way it webs through the body, up through the lower back, stunning and debilitating.

I noted, halfway down the hill, that we would be ending up at a different place than where we had begun. Carl's cheerful, sweat-laden response: "Yup."

"Excellent," I said.

Indeed, at the bottom of that trail, we had still two or three miles to go along a paved country road. We were a hop, skip, and jump-into-

Dublin-Lake away from our apartment. Fortunately, there's something about physical exertion—especially once back to flat land and pavement—that is addictive to me. It might be a struggle, but the adrenaline joy and sense of satisfaction that result bring to my mind the survival of the fittest. We, naked at birth, clawless, tailless, fangless *Homo s. sapiens* have survived as a species this long for a reason. We endure and we gain something from the experience. Lessons. For instance, when climbing a mountain, go step by step. Don't look too far ahead. Hydrate. And when you reach the top, take the time to breathe, to see the grand beauty and miracle of nature that surrounds you.

Learning the mountain's language has been a test of endurance for me. Most times, I only have a rudimentary understanding of it, but, occasionally, when I am not trying, I am as fluent in it as in the ocean's language. Maybe because, at some moments of a day—the foggy ones when purple haze lightly glazes the horizon—the hills are full of lakes and ponds made of gray and white and blue fog banks. The clouds are transient mountains, and the mountain becomes a silhouette of an ocean.

And then a gun shot from the gun club. Machine gun fire, rat-a-tat-tat, shatters the peace and I think of what this place might have been once, when people were killing people for big concepts like freedom and independence. And I imagine what it might become when Carl and I are no longer here: glaciers becoming water, flooding land, joining oceans and mountains. Climate disruption and wars will happen because people will be people. We learn lessons again and again because, too often, we aren't listening, reflecting, remembering this: that oceans and mountains speak their own vernacular, expressions of life everlasting, and that we have no control, no power, nothing but nothing as the wind howls and the breezes whisper through purple mountains and from sea to shining sea.

HOW WE GOT HERE

Growing up privileged—white, relatively wealthy, and American—I have had choices most people don't. And, as a Libra, I am ever balancing and weighing what's right, what's wrong. I had the choice to be here or there, to work or not to work. My questions were not "Can I afford this?" but the more puritanical "Do I deserve it?" or "Should I?" My choices, thus, have held a moral quality, a sense of responsibility. More often than not, they have resulted in consequences I had not anticipated. For example, attending boarding school. Not going to law school, opting instead to earn an MFA in writing. Marrying Carl. Moving to Jaffrey, New Hampshire, just for the winter, to be sure our off-grid system didn't freeze. These choices had all seemed so simple, logical. I might even say inevitable.

Were they? When I look back, along the long line of options given, decisions made, there does seem to be a connection, a shimmering filament that, like the stars in the sky, I can make into pictures of heroes and gods and dippers. Doesn't that sound like fun, to create one's own story, picking and choosing what to tell, pushing squares into circles so that it makes sense? And isn't it amazing to be of an age that one can smile back at one's youth and naiveté, remembering how one's imagination came up splat against the wall of reality?

By now you might be wondering, "Where are the chickens?" Patience. I didn't adopt them until I was fifty. I married Carl when I was verging on twenty-eight, the beginning of a fairy tale, if anyone believes in those. Riding off into the sunset, happily ever after.

But let's be real.

Once upon a time, in 1989, I was bartending at the Hot Club, a popular, waterfront bar in Providence, Rhode Island. An old generator building had been renovated into that small tavern, where the working world seemed to come on a Friday afternoon for drinks and hookups. The women bartenders were a no-nonsense group. They knew how to make a good cocktail and keep order when things got rowdy. Good money was to be made on a busy shift. Being new and, therefore, low in the pecking order, I was given the quiet shifts. Monday and Tuesday nights. Sunday afternoons. I worked the inside bar next to the grill that served hamburgers, hot dogs, and zucchini sticks. Customers would enter the single-story, glassed-in building and glance into the first bar room, where I'd stand chatting with a regular or the grill cook. I would smile in greeting. The customer/s would nod and wander off to the outside bar. It was nothing personal. People came for the water view. Mine was of the parking lot.

One Sunday afternoon, a zippy red Chevy Cavalier drove into that lot. I can still see it parked outside of the bar, and Carl stepping out of it with his happy, bouncing step. He—a Germanic blond, solidly built, cheerful fellow who knew both the customer at my bar and the grill chef—had come in to meet a friend for a drink before going out for sushi. His eyebrows were bushy and glowed white against his ruddy complexion. Soon enough his friend arrived and off they went. He returned a couple of hours later, ostensibly to get his car.

We had nothing in common. He played trombone in the R&B band Roomful of Blues and was on the road three hundred–plus days out of the year. That meant late-night partying, womanizing, and drugs, right? For my part, I had grown up listening to Roomful. I adored dancing to their music. So it was pretty cool to meet their trombonist. But I was an early-to-bed, health conscious, moody graduate student who liked routine and control. I found his sunny disposition and lack of stress, frankly, bizarre. Also, he smoked, strike one; he golfed, strike two; and he had a twenty-eight-foot King's Cruiser—a.k.a. wooden sailboat—named *SAGA*, strike three.

Simple? Logical? Inevitable? Three months later, I was at a Lake Tahoe ice cream shop in the basement of Harrah's Casino, where Roomful had a week's worth of gigs. Carl had just ordered a large bubble gum ice cream cone, a flavor he had tasted and said he didn't like.

"Why order ice cream you don't like?" I asked. *Sunny disposition alert*: he shrugged and grinned.

"I liked it as a kid. Maybe I'll learn to like it again."

Seriously? Eat an ice cream cone? Large? That he didn't like? Was he crazy? I had a history of anorexia. I knew the awful consequence of eating too much: self-loathing and the loss of any sense of control. Just the thought of eating food I didn't like rated a red-alert level of insanity that was only matched by Carl's bafflement about something called "dieting." He was a card-carrying member of the Clean Plate Club. He had no concept of restriction or discipline when it came to food. And I had no intention of being with someone who didn't at least attempt to maintain his boyish figure. Who didn't believe in exercise unless it was working outside, which wasn't an option when on the road. Who ate and drank with carefree abandon. Who smoked.

That was my first glimmer of what life with Carl would be like: more often than not, outside of my comfort zone.

We got married a year later.

How did I get from protesting to friends he wasn't my type to a lifetime commitment? I had always assumed I would marry—*if* I married—an intellectual professor-type like my father. Someone who would teach me to be who I was supposed to be, that perfect self I had been as a child. All that potential talent and intelligence, like a black star waiting to explode.

Instead of a knight-in-shining-armor living in an ivory tower of academia, like a moth attracted to flame, I fluttered around Carl. He did not live in the cerebral society of books or, most certainly, in the safety of routine. He did do what some people called living, thereby proving he had a higher tolerance for adrenaline rushes than I.

With him, though, my tolerance level rose. I could do things with him that I wouldn't do without him, like eat an ice cream cone or take off a day from exercise.

And, like my parents, he traveled. A lot. The movement and absence felt terribly familiar.

Whereas Carl and his troop of friends grew up together playing soccer and swimming in the summers, skiing and skating in the winters, and partying, I had led a more solitary life. Summers in Connecticut at one or another grandmother's, winters in Michigan. And then there were the family travels for my father's research. I had friends, of course. Each year, a best friend, but they were not forever because our family moved about so much. I don't remember the comings and goings. I do know that we would return from a summer in Connecticut, or a year in Vienna, and my old friends in Michigan would have moved on to new friends. What friendships I made in Stonington, too, came and went. Books, thus, became more reliable companions. I read. Nancy Drew mysteries and the Laura Ingalls Wilder chronicles. The Black Stallion series and *The Fellowship of the Ring*. *Watership Down*, *1984*, *Animal Farm*, *QB VII*, *The Andromeda Strain*. Wherever that moving target home might have been, I had my books.

The summer of 1976, I was thirteen and my best friend at the time was going to visit her father in Spain. Her parents were divorced, and so I was invited to join Julia until her mother arrived later in the summer. Julia's father lived on a *finca,* a country estate outside of Madrid. Julia's memories of visiting the place set off my imagination: we would ride horses, just like Alec Ramsey had ridden the Black Stallion and Ken had ridden Flicka. We would race through acres of fields and woods, our hair flying behind us, then swim in the lake. We would eat lots of delicious food yet stay slender because we'd be so active and busy. The joys and wonders we would experience! The explorations we would make! Best of all, I would become fluent in Spanish and begin my life of cosmopolitan, polyglot, intellectual itinerancy. I packed for the trip. A photograph of Julia and me just before we left shows us as two tall buxom girls, glowing in the sun. I am tan relative to Julia's fair skin, my light-brown hair shoulder-length and hers a blonde boyish cut.

We arrived in Barcelona and were deposited at her father's Spanish villa, as opposed to his *finca* outside of Madrid. I didn't know what had become of our horse rides and lake swimming. I'm not sure that Julia did, either. We entered from a narrow cobblestone street into a walled-in courtyard remarkable in its resemblance to a jungle. Endless, overgrown plants and an ancient tree with giant limbs overhanging creaky patio furniture. The garden was shadowed and cool. And the house? "Long in the tooth" is a phrase that comes to mind. Its bright yellow exterior had been blackened by the soot of time and pollution, and the interior evoked a dark, musty sensation, one cluttered with the accoutrements of a lifetime and the constant *olés* of bullfights that came out of the television in the living room where Julia's father sat, bundled up in a thick blanket.

Apparently, he was dying of cancer. He lived with his sister who seemed to dislike Julia and, by my association with her, me. Neither of them spoke English. I didn't speak Spanish. I was entirely dependent on Julia, and homesick. Each morning I woke with that leaden-heart feeling of emptiness and a hollow dread of the day. I entered survival mode. Activity and initiative were not options. To Julia's immense irritation, I preferred to stay at the house. There, at least, I felt safe. I knew what to expect. Routine gave me a sense of control. Hot cocoa with skin on the top and toast in the morning. Cold cuts and bullfights at night. Days, we were dropped off at a club to sit next to a swimming pool full of peeing, shouting children. Julia refused to swim and had no patience with my homesickness. Neither of us knew how to cope with our respective realities. Instead, she would go off and talk to acquaintances. I would sit lonely with a book. I read and wrote in a diary. The diary was my companion. It listened and understood, comforted me as I penned my obsessive ruminations of wanting to go home, and what was my problem? Why couldn't I just be happy?

The embarrassment of wanting. My shame, having to plead with Julia to call home. In retrospect, she probably wasn't allowed to use the phone to call the United States. In those days, cross-Atlantic calls were expensive. Julia didn't explain. She only refused, told me to grow up. Julia, who was coping with her own confusion and disappointment, her dying father.

We finally did call. My mother's voice—strong and vital—was a draught of hope. I wept, said that I wanted to come home. I, who had been so excited to travel. Who had chosen to take that adventure. She asked if I could wait it out another ten days until the arrival of Julia's mother, who couldn't come any sooner, and I couldn't leave Julia alone. There wasn't an option.

Ten days I stayed. Nine. Still homesick and pushed into uncomfortable situations. Eight. A Hitchcockian evening at a fairground hovers on the outskirts of my memory. Seven. As the day of my departure neared, a conviction that I would die. Six. I believed that I would never be home again. Five. I needed it so desperately. Four. I was powerless to get it. Three. The way there—a taxi ride through foreign streets, the airport maze, the eight-hour flight, customs—seemed insurmountable. Two. At last Julia's mother arrived. One. I could leave, and Julia's parting words to me as I left her were that she hoped my plane would crash.

It didn't. I made it home and nothing had changed except me, and so everything had changed. An ever-so-slight shift in perspective and home felt almost disappointing. Rather, I had disappointed. I believed I had failed, not lived up to or deserved such an exciting opportunity as a trip to Spain with a friend. Anyone else would have, at least, learned Spanish.

The next entry in my diary was written a month later, one month before my fourteenth birthday. I had arrived at Miss Porter's boarding school in Farmington, Connecticut, a freshman. My parents and I had agreed, months and months before, that it was a good idea. The theory: I would get a better education than the one I would have received at the public high school in Haslett, Michigan, where my best friends of years past had, one, gotten pregnant; two, threatened to beat me up, I have no memory of why; and three, acquired a boyfriend. I opted for a broadening of my life experience elsewhere. After all, my parents had both gone to boarding school. My sister, too. It made perfect sense. Certainly, my parents seemed to prefer the East Coast to the Midwest, where they felt closed off from the culture and energy of New York City, where they had met. And so I chose to attend boarding school. But even before my mother dropped me off there that first day, my heart compressed with a too-familiar sense of loss. Imminent. Eternal. Like death. I breathed shallowly on the drive there, in order not to fan the pain and dread and carsickness. We stopped en route for me to be sick on the side of the road. I reminded myself that this was for my own good.

We arrived at Miss Porter's and Ward dormitory. There we met my roommate, Alden, a cheerful, outgoing girl whose mother—a formal, coiffed lady—had attended the school. Alden knew all the New Girl–Old Girl traditions, the lyrics of the school song, and how to study, if not how to keep the room tidy. I knew none of the above. Alden invited me to join her for an opening day event. Unable to breathe through my panic, I opted to stay with my mother. Alden skipped off, blonde, pale-skinned, and bursting with anticipation because finally she would begin the lifelong promise of MPS: We come as girls, leave as women. That was to be aspired to because women are strong, smart, enduring. My mother helped me to unpack, then took me out for a chocolate ice cream cone. Before leaving, she urged me to participate, to go meet some nice girls at the soccer game. There, my wallet was stolen.

I reviewed the lessons learned in Spain and endured by counting the days until vacation. My mantra: I want to go home. My parents encouraged me to stay. After a month, they suggested I not call quite so often. I should try to immerse myself in the experience and to study. Not wanting to fail them, I tried.

Every morning I would wake up to the buzz of the alarm clock, my chest tight with black and breathless panic, my heart boxed in by anxiety. I lived in dread of the days, life, anything unexpected. I would attend classes but extracurricular activities existed outside of my capacity. I had no time for such things as sports and drama. I had to study. My initial fantasy of brilliance and extra credit had become a treading the waters of barely getting by. Piles of homework edged up my sense of panic, the challenge of existing. I tried to follow a strict schedule of what to do when. But how to control that monkey brain of mine that only found peace in obsessive rumination on home?

Like a cat purring to soothe itself, I counted the days on the calendar until I could go home. Over and over again. What might my parents be doing now? My cats? I would stare at the page of my biology book, read the paragraph before me, over and again, my ability to learn drowned in the rising tide of frustration and self-disgust. I used to be a good student, As and Bs. Now I got Cs, Ds, humiliating Fs. Whereas Alden learned her lessons without seeming to try, my concentration and focus dissolved into tears because vocabulary words and math theorems didn't stick in my brain. Latin declensions defeated me. *Amo, amas, amat, amamus, amatis, amant?* But maybe my parents didn't love me because they didn't say, "Come home." I wanted them to say, "Come home."

There were good days, and friendships formed. Water fights and brownie baking. Pizza orders and pranks. Crushes on two juniors just down the hallway from us. They were "old girls" who also suffered, at times, from homesickness—but how adult they appeared! They would comfort me as I wept. But my tears kept coming. Relentless, sucking. After a while, as my parents had done, they suggested I quit the self-pity and pull up my bootstraps. Try.

Walking back from a sandwich shop with Alden one day, she stopped to talk to some other girls. I needed to go back to the room to attempt, at least, to learn that damned Shakespeare sonnet. *To be or not to be?*

"I'm going home," I said. Heart-stopping words. I had not meant to say that. If the dormitory room were home, then what place had my parents in my life? By calling that room home, I had betrayed them. Never mind the fact that I didn't want that floral-wallpapered dormitory room to be home. I wanted home to be with Foggy and Bathsheba, and Mom's greyhound, Géza. Home was our Frank Lloyd Wright–style ranch house with its warm, radiant-heated floors and wall-to-ceiling windows to the south. The avocado tree that hung over the dining room table. My father's garden with its array of bearded iris, day lilies, and peonies. The rough-and-tumble lawn he loved to mow. And what of the twice-daily walks with my mother and Géza along the paths of the field in back of the house, watching flocks of birds filling the sky during migration season? These things were home, not that dorm room. Deep in my core, I rebelled against the shifting change I was undergoing, grasped to stop it, refused to adapt because that meant giving up the past, my parents, my home.

It happened anyway, along with a reworking of reality so my betrayal made sense. I didn't deserve to go home. I was a spoiled wretch. How dare I complain in the midst of such bounty? Opportunities surrounded me and I hadn't what it took to take advantage of them. I had loving parents who thought sending me to boarding school was the right thing to do. It was I who was wrong. I who wasn't good enough or smart enough. I who did not deserve all I had, pathetic creature that I was.

When I arrived home for Christmas vacation, I hardly recognized Foggy. He was more solid than the lithe creature I had left behind. My best friend now had a job *and* a boyfriend, and no time for me. My mother had pulled out her back and lay on the low-lying green couch. She needed my help, which meant she couldn't help me. As the holiday neared its end, my parents said I could stay home if I wanted to. They left the choice to me.

A hint of failure. Was I too weak to stick it out?

And to stay home meant a new beginning at that local public high school that I had previously shunned. It was co-ed. Boys scared me, and my ex-best friend's threat yet lurked. It felt harder to stay home than to return to Miss Porter's, which had, indeed, become a home of sorts, with friends with whom I shared my feelings of despair and absence, my failure to concentrate and, thus, my failed studies, my dark suspicion that something was wrong with me.

I went back to boarding school and kept counting. I counted the days until spring break, summer. And then I counted the calories I ate. The miles I ran. Nothing helped. Nothing held me. I felt not good enough because I always wanted. Forever needed. And so, hungry? Ignore it. Tired? Run farther. It was easy to ignore the feelings once the habit had developed. And still yearning for home. For four years.

My self split. Two parts, head and body divided so as to more easily ignore the pain of separation. To survive and make sense of that boarding school situation, I stopped listening to my needs and feelings because to recognize the emotional agony, to feel the anger and powerlessness of the situation would ruin the most important chapter in my story, the one about perfect parents, safety, and love: home.

People create stories to explain the world around them. For most of my life, this has been mine: my Spanish experience caused a fissure in my heart. My boarding school experience broke it. It was the "sink or swim" era, as one of my teachers described it years later. I believed I sank. For years, my story concluded with "if only." If only it could have been different, I would be the person I should have been: that sophisticated, polyglot intellectual who is able to concentrate, focus, accomplish. If only I had been stronger, more courageous, I would be a happier, smarter person, one more in line with Carl's playful *joie de vivre*. Instead, I seek home wherever I go, and leaving has a dread permanence in my heart because, too easily, it might be forever.

On our first date, Carl and I discussed marriage. Both scoffed at that useless institution. His parents had divorced when he was four. Mine were having a rocky time of it. And so when Carl asked me to marry him, I assumed he was joking. I said no. Soon enough, he recovered from that disappointment and asked me again, and again I said no. Patient and stubborn soul that he is, he asked me a third time. That time, my guard was down. He had gone out to play basketball with his buddies, and I had managed to work myself into a state of worry and neglect. Only a couple of months after the ice cream cone, I had, apparently, grown to depend on him and the relative steadiness he promised.

My family worried about my marrying a musician, with all the concomitant evils that suggested. All I knew was that I felt tolerable and acceptable when I was with Carl. And without him, not.

By 2006, Carl had left Roomful of Blues and was freelancing, producing albums, and no longer smoking, thanks to an acupuncturist. I had three unpublished books and numerous short stories boxed in a closet, and was a member of the governing board of Common Cause Rhode Island, a non-partisan organization promoting open and accountable government. We had moved into and renovated a glorious Victorian home on Angell Street in Providence and sold our first home on Cole Avenue. A tree had fallen on the neglected *SAGA*'s bow, and we'd given her to a young captain who intended to turn her into a racing boat. Carl's 1962 Bel Air had mice in its motor. My cat Agora had died. Kittens Nick and Nora had been adopted. Carl was turning fifty. Change was in the air and it was still a year before Barack Obama would begin his presidential campaign.

Change. Obvious to me now but not then, Carl and I thrive on change. Rather, we create it in the form of activity, movement that some might call chaos, with chaos being a state counter to that which I claim is my preferred *modus operandi*: reclusion, peace in the form of controlled, uncluttered order. I claimed to want routine and calm to the point of quietude, yet did seemingly everything possible to avoid that. When in Providence, if I wasn't writing, or attending a Common Cause Rhode Island board meeting, or working on the bounteous array of house and garden projects Carl and I came up with, I had an apparent compulsion to entertain, perhaps

Our new kittens, Nick and Nora

34

a genetic trait from the maternal side of my family (restaurateurs and liquor producers and distributors) and the paternal side (lots of dinner parties that prompted people to write thank-you letters for "such a delightful time"). Wherever it came from, Carl happily indulged my compulsion. And then there was the traveling to Stonington, Connecticut (my paternal ancestral family home), New York City (my mother and sister), and Peterborough, New Hampshire (Carl's family). Especially around the holidays, we would have debates as to whose family we would be with, and where to stay. The choices seemed never to include staying home.

One aspect of our vortex of activity had been a casual search for property near Peterborough, where Carl still had strong ties, not just family but scores of childhood friends. That search had picked up intensity ever since Sarah, Carl's niece, gave birth to her daughter, Peyton. Carl and I had jumped into the car as soon as we heard Sarah was in labor. She would be a single mother and had the full support of her family. Her parents, Tommy and Mary, would be there, along with her brother Jared and his wife, another Sarah. And so would we be. The two-hour drive from Providence to Peterborough passed in a sparkling of anticipation. Alas! We arrived at the hospital half an hour after the baby's swoosh and bob into a bathtub, our ever-so-slight disappointment for our tardiness alleviated by a vague sense of relief not to have been witness to the reality of Sarah's *passion*. And no harm done. We still had the pleasure of holding the itsy-bitsy baby, that bundle of energy swaddled in a white cotton blanket. I noted Carl's expression as he held Peyton: the slight dimpling and tremble of his chin that happens when he is deeply moved.

I, too, fell in love with Peyton. Feeling the heat emanating from her body connected me to her in a way I had never imagined. I fell to such an extent that, that afternoon, as we drove back to Providence, I debated whether Carl and I should have kids. I had never had an interest in having children. I was all for vicarious parenthood. But Peyton in my arms had opened up possibilities. A different way of life and being. Carl wasn't traveling as much. We could probably figure out how to be parents. Granted, the world was going to hell. And how selfish to add yet another person to the already burgeoning human population. A me blended with a Carl, and given our family histories, the poor kid would be bipolar. Not that that mattered. S/he'd be a good person. But think about the karma: I might have been a good kid, but Carl? It was a miracle he was alive to tell the stories of what he'd been up to. What if our kid decided to try drugs? What if s/he drove too fast? Skied, tripped, and fell? But how cute s/he would be. Probably a towhead with those impish eyes of Carl's and his freckles. Fair skin. A disposition toward skin cancers. Or maybe s/he would have the wild and disheveled eyebrows that both Carl and I contended with. My family's long fingers. A pianist!

As we approached the city of Providence, a full hour and forty-five minutes into that debate, during which time I realized helicopter mothers were amateurs when it came to the smothering and repressing I would wrought, I looked over at Carl. "Well?"

Carl startled.

"Well, what?" he asked.

"Don't you have anything to say about it?"

"You've been arguing both sides beautifully." He grinned.

I glared.

"It's your body," he shrugged. "I'm game either way."

That's one of the more bracing aspects of Carl: he, more often than not, leaves it to me, the ever-teetering Libra, to make the final decision. He, the Aquarian, goes with the flow. He can be content in most anything, so long as we are together. For better or worse, so long as I'm happy, he is happy. Maybe because he knows I'll end up doing what he wants to, if he really wants to.

As luck would have it, life swept us up again, and Peyton was there to distract us from a full-on commitment to parenthood. Who needed one of our own when we had Peyton to fill in the gap? She, from the nanosecond she was born, had a special connection: Carl was wrapped around her pinky finger. She could sneeze and he would be there in his entirely sporadic, devoted, there-in-spirit way. Thus, Peyton's arrival on this earth added that extra *soupçon* that motivated us to think more seriously about having a small, weekend place of our own in New Hampshire. A place we could take our cats. Friends questioned my rationale. Hadn't I said New Hampshire made me claustrophobic with all its trees, winding dirt roads, and lack of horizon? Why would I take on another place when I claimed I wanted stillness and calm, not more movement, stress, and responsibility?

Peyton and Carl taking a break from ice skating in NYC

My response: I was window shopping. I'm happy to look. Looking can't hurt, right?

The years passed. I got more familiar with the Peterborough area, the people. Day trips. Overnight visits. Our own form of couch surfing. With time, the ambivalence I felt about buying property in New Hampshire morphed into a desire to have a place there to call our own.

A month after Carl's fiftieth birthday bash—held coincidentally in Jaffrey, New Hampshire—a realtor acquaintance of his, who had attended the dance party, called to say he had "a sixty-four-acre lot off a Class VI road" for us. Apparently, we had mentioned to Andy that we were looking for property.

"Class VI," for the record, means not maintained by the town. And those sixty-four acres that used to be farm land with sheep and orchards had since grown back into woods, thick with brush and trees. Trees everywhere. Trees are wonderful. I love trees. I have been known to hug them.

But, as previously mentioned, I had grown up near the Connecticut seashore and, thus, was more accustomed to ocean views. There's a horizon. You can see out. You can breathe.

Of more concern, however, was that Class VI road.

I don't like to drive. I really don't like to drive at night. And I really, really don't like to drive in snow and ice, or along anything remotely narrow or crater-ish. And so, the Class VI road—what I considered, on a good day, a bucolic suggestion of a country lane—did not pass muster. From the very first time we drove out to see the property, as we bumped along that endless, rutted goat path, I had issues with it. We followed Andy's car, heaving and rocking deeper into the woods. I imagined driving that euphemism of a road every day. It seemed a rather challenging escape route in comparison to our house in Providence, where I could step outside the house and be somewhere. And how would we get out in the middle of winter? Dog sled?

A mile in, Andy parked his nice black sedan ever so slightly to the side of the unmaintained road. He got out of his car and pointed out "our" property. Its boundary marked by the meeting of two stone walls, it appeared to have lots and lots of flora, and a very steep incline. We exited our Subaru. Carl stepped forward, up and over the stone wall. Apparently, he was ready to tramp and bushwhack through that forested landscape. Andy followed, if with slightly less enthusiasm. In my imagination, he wears black slacks, an all-weather jacket, and loafers, which doesn't make sense, he being yet another burly, though clean-shaven, New Hampshire–raised fellow. He must have had on boots.

As for me, I stepped up onto the stone wall and down into the property, likely as not wearing the pair of brown leather hiking boots that I had bought years before and that had hardly any wear on them. I paused to take a thorny branch between my fingers, held it up so as to step through the offending bush, and proceeded uncertainly in their wake. We wended and wandered our way up that brushy, wooded hill, Carl and Andy waxing nostalgic about various classmates and teachers of yore. A half hour or so in, I approached Carl to point out that it was muggy and the blackflies were swarming me, if no one else.

"This might be a good place to build," Andy suggested from where he stood under a looming pine tree. I narrowed my eyes at him to better study his square head, Buster Brown hair cut, and pale complexion. Whose side was he on, anyway, with his ever-so-slight New Hampsha twang? He pointed out the relative flatness of the area, the possible eastern and southern exposures. Carl and he had a brief *tête-à-tête*, orienting themselves to where we were relative to the rest of the world, finally agreeing that there was likely an eastern view of the entire Wapack Range. I swiped at blackflies, clueless of what that might be. I envisioned wagon trains, and determined this was a sales pitch on Andy's part. All I could see were masses of trees and bug bite welts, and Carl, who glowed with anticipation. His blond brows looked particularly unruly and he was covered with sweat and plant detritus, as if he had been mowing leaves. A dead mosquito lay flattened on the side of his ruddy neck.

"It's a beautiful piece of property," Carl said, nodding.

"Once you get here," I replied, unmasking again our glaring difference. I have a Rhode Island perspective. A drive from the East Side of Providence to the village of Edgewood, a distance of five miles, is the New Yorker's equivalent of driving to New Jersey. It is a journey to be prepared for with a daypack and a first-aid kit. A snowflake falls from the sky? I am home with a fire in the fireplace, cooking soups and baking brownies and bread, braced for a blizzard and with sufficient groceries to feed ten half-frozen, starving waifs. I encourage the concept of "snowed in" for its holiday-ish, snug as a bug feeling.

Carl, on the other hand, likes to drive in blizzards. He goes out in the snow for the joy of it, plowing his Subaru into five-foot snow banks with a whoop and a holler, thereby exposing his New Hampshire roots. All his New Hampshire friends like to drive in snow. It's a point of pride to a New Hampshirite not to be fazed by a foot or two of that white, fluffy precipitation. There is no such thing as "snowed in" in New Hampshire.

As we drove away from the land, back along the "road" to relative civilization, we considered the pros and cons of that beautiful-property-once-you-get-there. I convinced myself that it was the shameful prima donna in me, not common sense, suggesting that I run away from that sixty-four acre lot. I ignored my sympathetic system's response, also known as doubt. Thereby and once again that long-ago dissociation of mind and body, developed to keep me feeling safe and above the sloshing rapids of change, worked its magic. I didn't think—or did I?—that the trail Carl blazed, and

I followed, directed me toward a life for which I was dubiously prepared. I didn't suspect that the life in New Hampshire we were contemplating would be an adventure for us both but a risk for me. Carl would be coming home. I would be leaving mine. He was falling in love with the land. I was in a danger zone, full of unfamiliar situations and tentative friendships that depended on Carl.

But why would I think about any of that? We weren't moving there! We were only considering buying a piece of property that I would never have bought by myself.

Neither would I have spent hours sanding and shellacking the brightwork of a wooden boat.

Nor would I have learned to ski.

Nor would I have done any of the other hundreds of things that Carl has dared me to do that I have enjoyed in retrospect, if not during. I step off cliffs daily because I know he is there to catch me. I might not have been ready, but I somehow trusted that with time, I would fall in love with that land, too.

Like a perfectly ripe peach falling from the tree, we fell. After years of thinking about buying something up in the hinterlands of southern New Hampshire, it was time. We made an offer on the lot located at the end of the Class VI road in Jaffrey. There, someday, we might build a small, off-grid cabin in the woods. A weekend getaway. A holiday roost. Simple.

Until, not long after we bought our land, Jaffrey prohibited building on Class VI roads. To build, we would have to upgrade that goat path to Class V level, which would include moving large boulders, doubling the width of the road, and possibly creating a lifelong enmity with our new neighbors who lived on that road for the very specific reason that it gave them privacy.

Fortunately, or unfortunately—as ever depending on mood and interpretation—we had, in the meantime, hired a forester to write up a forestry plan for our property. He walked us through it with assurance and clarity. He noted the various tree species and described the ecology and history of the land gleaned from that information. By the time we stopped at the top of the hill, near the stone wall that was the western boundary between

View of Mount Monadnock from Parker Pond

us and the next lot—129 acres of land abutting our 64 acres, inclusive of a 34-acre pond and a 5-acre field that had *two hundred feet of Class V road frontage,* the forester had gathered all the information he needed. He was a man who could see through the trees.

"You'll have a nice view of the Wapack Range and Massachusetts to your east and south. But that's your real view." He nodded west, across the stone wall at . . . more woods. "There's Mount Monadnock. Too bad your neighbors own the trees. Let's take a walk."

And over the wall he went. We followed. Why not? We were in New Hampshire. Live free or die. It wasn't hunting season . . . or was it? We wore no orange. But I was with Carl and the New Hampshire equivalent of Strider from *The Fellowship of the Rings.* We'd be fine.

I kept my eyes peeled for Orcs as we walked down the steep hill, across a swampy area, and continued along a logging path that our Strider said was about ten-to-fifteen years old, judging by the size of the undergrowth and trees. Stepping over a stone wall that had once bordered a sheep pasture, he continued to assess the land. He pointed out coyote dens, moose poop, moss to the north. He noted that the trees—oak, maple, a few

41

cherry, hemlock—were older and of better quality than those on our piece of land, which was fast becoming less interesting. Signs of beaver activity. A hemlock grove with a path that led us along a berm that circuited the aforementioned pond. Prehistoric. Eden-like with its water view of Mount Monadnock.

I noted the surfacing of Carl's five-year-old self, freckled and joyous as he marched through the hemlock grove. Excited. Scheming. Definitely imagining himself younger, because he kept talking about cutting down the trees ourselves. Clearing the land ourselves. Building the house ourselves. Farming. Apparently, he was channeling Scott and Helen Nearing. I saw us more like Lisa and Oliver Douglas of the sitcom series *Green Acres*.

I have wielded a wonder bar, knocked out walls, lugged garbage cans full of horsehair plaster down three flights of stairs and sheetrock up them, once in a skirt and heels. All satisfactory work. Just not something I would choose to do, given my preference for pushing a pen, not a wheelbarrow. Left to myself, I prefer to write, to read. Stillness and calm are good. If I could only get myself to sit still and be calm. As a friend once pointed out, Carl and I have trouble *being*. We are always *doing*. Whirling dervishes, flinging about four states, sometimes together, sometimes passing like two ships in the night. We are doers. Creators of homes. And I am a sucker for Carl when he exudes that happy, boyish spirit and its pull to a life that I fear. How could I say no to that happy man who had lost his heart to, let's count them, 193 acres of land?

N. O. No.

Too late. Carl and, yes, I had lost all rationality. The question we asked was not one of financial feasibility. Nor did we stop to create a logical, step-by-step master plan for our future. Pure and simple, we looked at each other and wondered how we might buy that second lot.

In our rewriting of history, of course, we point to the *fact* that, because Class VI roads in Jaffrey were closed to building, we had *no choice* but to find an alternative entrance to the property via a Class V road, *ergo* we *had* to buy the second lot. The consequence of that *analysis?* By 2007, we owned both lots, and life had become more stimulating, if for no other reason than this: the concept "off-grid" raised eyebrows at the Jaffrey Town Hall.

Granted, everything we did raised a brow. "Natural selection," "survival of the fittest," and our own version of Swamp Yankee stubbornness were phrases that came to mind during our year-long process of putting the land into a conservation easement, logging the lots, and permitting and building the logging road that would be a driveway some day but not this day. Thus, as previously mentioned, the name Darwin's View. Everything to do with the place required an adaptation of reason to reality and a suspension of disbelief that we were going forward when logic and budgetary concerns would have stopped most people. But forward we went, because the name Darwin's View gave the land a sense of significance and purpose. It was bigger than just us. Like the surrounding mountains, this land would be not owned but stewarded. We would not possess it, we would be possessed by it. The land demanded something of us. We had only to figure out what that would be.

Five years passed. I was writing, playing the flute, and continuing my education in Rhode Island politics. Carl gigged and produced records. We threw parties; replaced our water-lusting lawn with utilitarian, vegetable-growing raised garden beds; and contemplated the kitchen of our East Side of Providence Victorian home. We had bought and completely renovated that house in 1998–99, but had not succeeded in making the kitchen workable to our more open-plan, open-flow lifestyle. The back staircase—which led down to the side entrance of the house and the basement, and up to the second and third floors—was in the way of the area that should have been a sunny breakfast nook. The galley kitchen was a perfect working triangle, and its cupboard pantry was

Our Angell Street Victorian

one I would try to recreate, unsuccessfully, in years to come. But the fact remained, the kitchen was cut off from the dining room and living room by walls. In our initial renovation, wonder bars in hand, we had debated tearing down those walls, but cost, time, and the main chimney of the house had thwarted that endeavor. That was then. This was now. Time ticked and Carl and I would, more and more often, consider how we might change things up a bit.

Maybe build an underground garage that was connected to the basement. That would provide us with more garden space and open up the possibility of bumping out the kitchen over the current driveway. But no. The kitchen, though enlarged, would still not allow for that sunny breakfast nook because there, in the midst of everything kitchen, was the staircase.

Maybe we needed to remove it. But how get out of the basement without a staircase? Exiting the basement was important, especially if we were now entering the house via an underground garage.

What about an elevator? Replacing the staircase with an elevator would work. Wouldn't it be cool if the elevator went up and down using some sort of water displacement system? A gray water system, the circulation of which aerated plants that cleaned the water? Mother Nature does it. Why couldn't we? And how beautiful would that be, an elevator made up of water and plants?

That was a crazy, complicated plan, which did not necessarily discount the idea, because that shiny concept began to address the biggest problem of all: of the many rooms in that beautiful house, only a few were accessible to my mother, who had recently been diagnosed with Parkinson's disease. She lived in New York City but would visit, and we wanted her to continue to do so. Thus, the elevator was not as absurd as it might initially have seemed. And dreams, like window shopping, don't cost anything. That was proven by the fact that ever in the background of these Providence-oriented ideas percolated the possibility of building a place on our property in New Hampshire. It was a low but constant undercurrent, and began to distract us from our Providence kitchen concerns.

Our intention was to build a small place on our land and to go about it in an informed, process-oriented way. Early on in 2007, we held a charette on the newly cleared hill. We invited an energy consultant, Margaret Dillon, for her expertise in envelope and whole building systems; and a mechanical engineering consultant, Henry Spindler, for his input on possible heating and distribution systems. Both Margaret and Henry would help to create an integrated design process with an architect, also invited, and our builder, Tim Groesbeck, a childhood friend of Carl's. We also included in the charette a landscape designer friend from Providence, Mary Dennis, who would help us with the overall aesthetic. Everyone arrived via our impressively built, curvaceous, it-would-be-a-perfect-driveway logging road. With a gentle breeze blowing, electrical loads were computed along with snow loads, and the term *energy budgeting* was bandied about along with the question, "Will this be a full-time or part-time house?"

"We don't know," Carl and I would respond.

"A lot depends on that," Henry noted. "How much water and effluence flows through the system—."

"We don't know." I smiled and shrugged. "Most likely part time."

"But we might come up here for longer periods of time," Carl added.

"No need to cut off our noses to spite our faces," I agreed, because who knew? "We want it to work no matter what."

At the end of the day, the architect showed us his portfolio. It, most unfortunately, included photos of one of his beautiful homes burning to the ground days before it would be finished. We decided to interview other architects. They came and went, carrying with them the marriage of our ideas and their programming, also known as plans, all of which conflicted with our budget. The challenge of finding a balance between the two kept us driving back and forth, Rhode Island to New Hampshire, with an occasional detour down to New York City or Connecticut.

How would we pay for the New Hampshire house? From the kitchen layout and my mother's visits to the vague sense of having too much stuff, Carl's and my conversations had begun to morph from "let's renovate this place" to "let's sell it." An act I rebelled against. The Angell Street house was my forever home, and the neighborhood was full of friends and familiar haunts. I had always said I would never move out of that Victorian house. Certainly, I would never give it up to move to New Hampshire. No!

Months passed. Building some sort of structure up in New Hampshire distracted us more and more. Carl became enamored with slipform stone masonry, a brief fascination that blessedly hit the wall of reality when we went to visit the fellow who wrote the book on it. Conveniently, he lived in the Monadnock region. That kind sir gave us a tour of his house, going through the process of how slipform stone masonry works: First, build the forms. Place the forms where you want the wall. Fill the forms with stones and rocks. Pour in cement slurry. Let the wall dry. Remove the forms and put them on top of the wall. Repeat.

It's a labor-intensive job, moving the stones, stirring the cement slurry, lifting the stones up, and lifting them higher. At the end of the house tour, the builder studied Carl and me. He noted that he and his wife had been young when they built their house. He offered this point up not as an insult, but as a reality check. The physical aspect of the house building needed to be a major consideration in our decision-making process. I made sure we considered it. And, thus, with time, we moved on. Though only so far as to include the question of how to pay for whatever we built with the help of other people. In window-shopping fashion, Carl and I came up with the brilliant idea that selling our forever home in Providence and replacing it with a smaller place made sense. With what money was left, we would build our off-grid cabin in the woods.

Fascinating, isn't it, how talking about something, familiarizing oneself with the idea, considering it from all its angles, allows for the inconceivable to percolate down into one's heart until one day the impossible has become inevitable?

We began to look at houses that were for sale on the East Side of Providence, with an occasional dip into the neighboring areas of Pawtucket, Pawtuxet Village, and Edgewood, the latter three being located just outside of Providence city limits. After weeks and weeks, and one home inspection that left us running away from a deal, we gave up. On the lovely summer evening of that failed house inspection, we sat on our outdoor wraparound porch and looked at our blooming clerodendrum tree, a Florida plant, zone ten, that was not supposed to be able to thrive in zone five in New England, but there it was thriving. So were the peonies and bearded irises. And closer at hand, in the shaded area, our Jack Frost brunneras, my grandmother's bleeding hearts, and a variety of hostas all exhibited our years of experimental gardening and *ad hoc* creation of home. Birds sang in celebration of our reprieve and our neighbors' towering, swaying pines whispered relief. We opened a bottle of red zinfandel and toasted our near miss of buying a (much) smaller house that was sandwiched between the aromas of an Indian restaurant and those of an auto shop.

Over puttanesca pasta, we sensibly decided that we would stay in our home through the winter, during which time we would declutter, pack, and prepare the house for sale. In the spring, we would put it on the market and find a place to rent until we discovered a suitable house to buy on the East Side. Always on the East Side, where we had both lived for our entire adult lives, initially apart and now together. If the Angell Street house wasn't, the East Side was our forever home.

A week or two later, Carl was at a gig, and I had settled into bed with a book. The phone rang: It was Deej calling to say that he had been walking Jack, his beloved mutt terrier, who had a favorite sniff-and-pause place on the lawn of the big-windowed stone ranch house that Carl and I had pointed out as being interesting. We drove by it whenever Deej and his wife, Kyle, had us over for dinner. In the Edgewood area of Cranston, right on Roger Williams Park, the stone house stepped back and away from traffic. Hidden in the shadow of a towering magnolia tree and sporting huge picture windows as walls, it looked remarkably like the Frank Lloyd Wright–style home of my childhood. Deej said it had a For Sale sign in front of it.

Directly after hanging up the phone with Deej, I called our real estate agent to make an appointment to see the house.

It has been pointed out to me that it is not executive function that I lack, but the ability to anticipate consequences.

But the stone house made sense. The beauty of the park right across the street. Deej and Kyle lived a block away. Sure, it was in Edgewood, where Providence meets Cranston, thus not at all walking distance to our favorite grocery store and restaurants. But we loved the house. Those big windows that looked north, not south, thus no passive solar heat. Lots of big, expensive windows leaking fossil-fueled heat. The house was not twenty-first century tight but mid-twentieth century wallowing-in-cheap-oil space. It had a big lawn that I would get rid of in favor of raised garden beds. The basement was huge and high-ceilinged, the perfect man cave. Although I had said to Carl I had only one more house in me to renovate and that one would have to be in New Hampshire, this one was special. We made an offer. The sellers accepted. We had ourselves a deal.

The Beachmont Avenue house in Edgewood

And a house with a brand-new roof that had to be replaced within a year of purchase due to leaks, plural, and floors with a bizarre tilt because the entire foundation had sunk when the house was built in the 1950s. We'd hired three engineers to confirm the stability of the structure prior to buying. All of them assured us that any settling had happened years before. That funky house had a fifties-style wet bar and Vitrolite tiles in the bathrooms and kitchen that didn't line up due to the fallen foundation. Not a problem. We would end up removing all those tiles in subsequent renovations. As for the 2,000-gallon underground oil tank, that passed inspection prior to the sale but, two months after the sale—when our insurance company canceled us because we had a 2,000-gallon underground oil tank and no insurance company would cover us unless we dug it up—that tank was found to have huge holes in its bottom. The Rhode Island Department of Environmental Management, thus, facilitated our initial meeting with our new neighbor.

In the midst of renovations, we give parties to get the feel of a place.

48

We knew none of this yet. All we knew was that we had ourselves a project. The day we closed on the new house—remarkably, around the same time as the tidal wave reverberations of the loud POP! of the 2008 real estate bubble—we applied for and got a building permit to renovate the kitchen. We arrived at our new house with our wonder bars and hammers. That day, we took out the wall between the kitchen and the dining/living room. Thereby began a whole house renovation.

Activity is an excellent way to avoid thinking about consequences or feeling tectonic shifts of change at a primal level. Providence held my identity. It was Goldilocks-just-right for me. Not too big, not too small. Walking, not driving. Familiar places and faces that I had seen in passing for over two decades. That kind of familiarity creates a web beneath one, a support and coziness. After we bought the Edgewood house, Carl and I told ourselves we would still live in Providence. Our new address proved it: the house was in Providence because the city had bought all the property surrounding Roger Williams Park in order to control what was done on its edges. But as we drove back and forth, Providence to Edgewood, every morning and night, to work on the new house, we tended to be rather quiet. And every so often, especially when we got stuck in traffic, I would look over at Carl and ask some rendition of, "Is this a big mistake?"

Too often, he would look as worried as I felt. But then the traffic would start to move, and likely as not, we were late to meet an electrician or plumber, or friends for dinner. Walls came down. Kitchen cabinets, knob-and-tube wiring, and old plumbing were ripped out, and so, too, my heart. I was deep in the mire of not wanting to move from Angell Street, never mind renovate a house in Edgewood. I wanted to stay in our Angell Street home with its subtle sesame and sage gray walls. Peace and stillness. My childhood cognitive dissonance was back and in full throttle: "This is entirely wrong. I don't want to be doing this! I want to go back!" opposed "I made the choice. Time to deal with the consequences. And stop weeping! Some people are homeless. You have two roofs over your head. What have you to complain about?"

The renovations proceeded. So did the preparations for selling the Angell Street house. I busied myself reverse-acquisitioning—books, clothes, kitchen paraphernalia, let it all go! Carl, in the meantime, packed paper-clips and rubber bands, his collection of *Trombone* and *Wooden Boat* magazines. Twenty years into our marriage, I discovered that he is a true "waste not, want not" Yankee. But who had time to contemplate that?

Too many balls in the air, yet we carried on. Our life was a NASCAR race and we were the drivers, in control in a seat-of-our-pants way. We had a plan, a "two-for-one" deal. According to our *modus operandi*, that meant hiring an(other) architect to draw up plans for a New Hampshire house at the same time as we renovated the Edgewood house. The year 2011 consisted of a lot of plaster dust in Edgewood, low-grade panic because the Angell Street house wasn't selling, and weekly drives to New Hampshire to meet with said architect. We changed real estate agents, dramatically lowered the price of our Angell Street home, and moved into the new house. A young couple, Katie and David, who lived on a boat yet liked warmth on a cold winter night, agreed to housesit our Angell Street home until it sold, which it did at the end of the year. We could finally settle into the new house, our Beachmont Avenue house, and how appropriate a name because it was, indeed, between the beach (my family house) and the mountain (Jaffrey). Beachmont would be our transition home.

In February, 2012, our then rendition of an architect presented us with the final building documents for the New Hampshire cabin, the fruition of our enormous carbon footprint and his massaging of numbers.

How did we feel? Disappointed. Those plans didn't include our bigger-than-us ideas. No big windows oriented to the south for passive heating. No radiant heated floors. And what of the greenhouse? We had hoped to grow greens in the winter, my tendency to "bury" seeds notwithstanding. The greenhouse would have been connected to gray water/sewage treatment ponds that would also have served as rainwater catchments from the roof to water the gardens, with all connected to a chicken coop (?!) . . . somehow. We didn't know how, but they weren't in the house plans, those dreams of ours that everything would be part of a whole, each aspect fitting into the others. A microcosm of that delicate and precious balance, life.

And then we showed Tim the plans. A no-bullshit kind of guy, known for his Yosemite Sam temperament and Energizer Bunny work ethic, he took one look at them and scoffed.

"You can't afford this. It's three times your budget."

Over beers, we discussed our options. Tim suggested we put up a timber frame house. Simple. Two-and-a-half months later, he was cutting the timbers for that 24-by-38-foot structure. On April 20, 2012, our foundation was poured. April 24, the deck was put on the foundation. The timbers were framed up and ready to go. May 1 would be the "barn raising" of the structure that would be our weekend cabin.

The night before, Carl and I stayed at friends' B&B in Dublin. Rising early the next morning, we arrived at Darwin's View at 7:30. Tim and his crew were already there, appropriately dressed in all-weather gear. Carl and I, dressed in sneakers and jeans, wondered what we could have been thinking. Darwin's View and Mount Monadnock had a history of tossing foul weather at us whenever we had parties and campouts. What else but drizzle, rain, and mid-forties temperatures could we have expected?

Blackflies. It was May and there was not one in sight. And, miraculously, no wind. Carl might have been purple with cold by the end of the morning, but the weather couldn't have been more perfect. Twelve-year-old Peyton huddled next to me under an umbrella. We watched the second bay of timber rise up in the air, floating through the skies and down, fitting perfectly into the holes in the floor. She looked up at me and smiled her snaggletoothed smile.

"I didn't think this would EVER happen."

Peyton with umbrella

May 1, 2012,
the "Barn Raising"

The timber frame

Carl and I can be so far on the edge of the cutting edge that we call it, fondly, the bleeding edge. And I sometimes wonder if it would have made a difference if we had known then what we know now. Probably not. Because every step of the way we took the most rational, fiscally responsible, executive-functioning route we could possibly take.

"We," of course, being the wild card factor in that sentence.

After the raising of that beautiful structure, Carl and I were charged up and ready to go. Tossing caution to the wind, we swooped down the hill we had been trudging up for six years, so caught up in the excitement of another project that we didn't note the sharp swerve of our life's course. We had decisions to make. Who had time to consider consequences? We had originally intended to listen to Julie and just put up the structure, camp out for a year or two, get a sense of the place, yet when faced by certain questions—Did we want finished floors? A bathroom? Interior trim and doors?—it seemed practically profligate to stop. We had Tim and his crew available and on-site. Moving all that equipment off-site only to have it come back again would be a waste of time and money. And windows are good, we agreed. Lots of energy-efficient windows. A solar array fit into our off-grid plan, if only because getting wires up to the house would cost more than buying storage batteries. And all those tax incentives and rebates. It's as if you're getting the thing for free, if not for the fact a major outlay of cold cash is required up front.

It was somewhere in the midst of all this activity that we got a call from Julie. We were running errands and so in the car. I answered with boisterous cheer, because I love to talk with Julie. She is so sensible and clear. But this time her somber voice caused me to motion to Carl to pull to the side of the road so that we might be fully present while she informed us that what we had done that had been so exciting and well-thought-out and fiscally responsible had, in fact, put us in a hole we would not be climbing out of any time soon. Having listened to her all-too-apt and cogent numbers, I suggested we could sell the land. A deafening silence from both Carl and Julie followed that proposal. And then Julie informed me that we could never sell the land. We owned it and would own it forever because we would never get out of it what we had put in.

Given my personality, one might think that would be reassuring. To know I have to keep something. That it is there for me *forever*.

That word *forever* shimmered into my consciousness, like a word in a Magic 8 fortune ball, slowly rising up through the liquid to express panic and regret. What had I done? Not the we of Carl and me but the I inside me. Was this what Me, Myself, and I wanted? Was I headed in the right direction? After all, if Carl weren't around, would I live there alone?

No.

It was right around that time that Carl accepted a gig with Jimmy Vaughan and the Tilt-a-Whirl band. The band was based in Texas. They traveled. A lot.

CHICKEN REVERIE TO REALITY

I could and, most certainly, have on occasion blamed our chickens, specifically our Rhode Island Red cock, Big Red, for what I, in moments of perceived powerlessness and rejection of all responsibility and free will, have called our *forced* move to Darwin's View. I feel justified in doing so because it is true. Our chickens have had more influence on our lives than even I care to admit.

In the spring of 2012, Carl returned to life on the road. We had just signed off on the architectural plans for the house in Jaffrey and looked forward to the building process. I was wallowing contentedly—if ever I could be content—in our newly renovated Beachmont house. After a full year of renovation, we had settled in. Months had gone by, and it had taken on the qualities that had been Angell Street's: the warmth and comfort of home, with all its trappings, not least, friends. I nested there, cozy in my office with its wood-burning stove and wood paneling. From the safe distance of my office, I wondered if those were weeds sprouting up in our raised garden beds. A sparkle of my schedule threatened to be available for calm. With an intake of a breath, on the verge of *om*, I noted how much time I was spending in the egg section of the grocery store, debating which to buy. Local? Organic? Free-range? Those terms were suspect. I imagined that agro-business marketers, whose jobs depended on profits, not ethics, defined those terms loosely.

One might ask, Why now? I had been a pescavore, eating fish, eggs, and dairy but not land meat, since I was seventeen. Why the sudden concern for all those miserable eggs? For those poor chickens clumped together in battery cages with no room to flap their wings or peck and scratch, stressed out and claustrophobic and debeaked? These images began to haunt my waking hours. My debate in the egg section took longer and longer. People have no respect for chickens, calling them dirty and sadistic. I wondered. Wasn't calling them dirty and sadistic putting some arbitrary human morality onto creatures that have their own set of rules? Chicken rules. Spend time with chickens in happy circumstances and you will find that they do have rules. Tough love that seems unfair to my eyes, pecking and chasing the lowliest, but it's about survival, not fairness. And these days, *Homo s. sapiens* aren't ones to point fingers when it comes to taking care of the weak and poor.

The point being, I wanted to eat happy eggs, which meant happy chickens, and the only way to know they were happy would be by seeing them and their living conditions with my own eyes. *A posteriori* knowledge: I would be taking care of them. In charge.

I began to surf—Carl might say obsessively—the MacMurray and My-PetChicken websites. I perused catalogues on mail-order fowl and books on chicken care. Carl became more than a little concerned when he noticed me petting photos of chicks in the magazines. He questioned whether this was really the time to get chickens. Still under the influence of those cute chick photos, I asked, wasn't it better to take on the responsibility for six chickens than to eat eggs laid in morally challenged conditions? Our chickens would represent our hopes for a better, more compassionate future. Some people have children. We would have chicks. Besides, if not now, we might never have chickens.

That being exactly Carl's point. But he wasn't there to reiterate it. He was on the road, leaving my worries to fester, and my diagnosed ADD brain to turn obsessive. I debated, as usual, with myself. Carl, as he had with the "let's have children" debate and, perhaps, assuming I would come to a similar conclusion, left it to me to decide. The determining factor? Two girls who lived down the way from our Beachmont house, and their parents, committed to help with chicken care. With their *imprimatur*, I put my money, time, and sanity on the line: I ordered six chicks, the maximum allowed in the city, hens only. A mix of breeds.

Tick, tock. The May 1 "barn raising" in New Hampshire of what would be our "weekend" "cabin." Family visits to New York City and Stonington, Connecticut. Choosing bathroom tiles and light fixtures for the cabin house. Carl on the road. In between Common Causing, writing, and fluting, I read up on all things chicken and bought a heat lamp, chick crumble and grit, and a waterer.

Fully prepared for their arrival, I worried. Late at night on into the wee hours of the morning, I ruminated that maybe owning chickens wasn't the most brain-trust idea. This being the point Carl had tried to make. But countering my doubts was the fact that I agreed with all those who have said that a society can be measured by how it treats its animals. I determined that, with our chicks, I would balance out some of the bad karma of American society's heinous factory farms. Which included the hatching and shipping of chicks.

What had I been thinking, perpetuating such a business? Countless chicks born, and half of them dumped into barrels to suffocate or sent through a chopping machine to be put into cat and dog food because everyone wants happy eggs not cock-a-doodle-doos, pullets not cockerels. I was no different and noted the contradiction between what I said and what I did. I was a hypocrite, this not being the first or the last time I would claim that title. Because I might not know that I know what I know, but I do it anyway. Against any and all good advice.

Our newly hatched chicks shipped out. They existed. So did stress-over-drive. I still hadn't recovered from Edwina being taken to "the chop" in *Chicken Run*, and all my friends warned me of the inevitable slaughter. Hawks. Fox. Fishers. Carl had promised to build a Fort Knox coop, but, ahem, he was on the road. He left a week before the chicks arrived. Chicago. South Bend. Auburn. DC.

The good news? Everyone I spoke with had the same solution. The trick to surviving the death and destruction, the haunting sight of blood and feathers everywhere, in short, the heartbreak was *to not name them.*

The chicks arrived by USPS on June 26, 2012, in a cardboard box with holes in it so they might breathe. A pile of variegated balls of fluff who, with time, would distinguish themselves as one Dominique, one Australorp, one Bantam Buff Brahma, one Silver-laced Wyandotte, one Cochin Partridge, and one Rhode Island Red. Their alternative monikers being Ping, Panda, Lola, Chipper, CooLots, and Rhoda Red. Because I might have been warned about their inevitable demise, but these wee peeps would be my pets. Naming them confirmed that status, as did setting up their chick coop in the bathroom off my office.

I had prepared a large plastic kitty litter box with clean pine shavings, the waterer full of water, some chick grit, and chick crumble. A red halogen lamp balanced above, clamped onto the sink's towel rack. As directed by all things Google, I had turned on the lamp to warm the box a couple of hours before our post officer typically arrived. Ding, dong! I called Kyle, who came over to celebrate with me. So did the two girls. They watched as I carefully lifted each chick out of the delivery box and set her into her new home. The chicks cheeped and wobbled adorably, warmed by the heating lamp. They teetered about, tipping over the food, pooing in the water, and looking irresistibly cute.

At which point, and to my horror, the girls reached in to grab their respective chicks. The chicks, in turn, ran away, chirping loudly in protest. The girls persisted and successfully grabbed hold of their peeps, Lola in one fist, CooLots in another. I watched, breathless, and hoped the chicks wouldn't be crushed by such tender, loving care. Kyle did not seem to be alarmed, and so maybe I was being overly protective. Even so, I suggested that perhaps the chicks were stressed and needed to be left alone. The girls ignored me. And then Lola struggled free and fell the two feet or more to the ground. Durable creature that she was, she stood, dazed and shaken. I picked her up, set her carefully into the chick coop, and suggested it was bedtime for the chicks.

The girls left. Kyle and I called Carl, who was in a hotel room somewhere. We sent him videos and photos and it all felt familiar, him on the road and me at home, facing the consequences of my choices.

Over the next weeks, those personalities proved themselves to be a motley crew of activity, skittering about, testing their boundaries, their tiny wings. I tidied their pasty butts. Fed them. Hovered and worried. They grew, fluff making way for feathers. Ping earned her name the day her practice flapping ended her up on the wrong side of the coop box. Her gleeful announcement of escape swiftly turned into the alarm of separation anxiety when she realized no one had followed her. Lola, a bantam and thus half the size of the others, resembled the cartoon character Speedy Gonzales. Feisty, she would dash between her siblings to snatch a treat and then race away, dodging pecks and typically ending up huddled under Rhoda Red, who was the largest of the chicks and protective, though not as motherly as Panda. Even at a young age, Panda had a gentle quality about her, as if guiding the others, chiding them when they got too rambunctious or aggressive. Chipper, the intellectual of the bunch, would stop suddenly to stare into space, perhaps contemplating one or another of the worries emitted by CooLots, who, as her legs grew feathers,

The first chicks, growing up in the bathroom

emitted a constant stream of clucks and trills that sounded remarkably like "oooooh no, oooooh no."

The chicks' cheeps, initially endearing, soon began to grate on my nerves. If I couldn't hear them, they had obviously died. If I could, it was nails on a chalkboard. They were, after all, calling for their mother. I had no feathers or fluff to keep them warm and safe. All I could do was tend to their kitty litter pan with its wood chips, chick feed, and water, and make sure the heat lamp kept them neither too hot nor too cold but just right.

They quickly outgrew that kitty litter pan. After one week, I moved them into a big cardboard box. After three weeks, I created a Styrofoam board barrier with a large branch onto which they flew, week four. By week five, their realm included the entire bathroom. The word "sanitary" echoed in my brain. If I wasn't careful, they would take over the house. It was too late to mail-order a coop. Would Carl ever have the time to finish the coop he had begun?

The two girls, who had hoped to help with the chicks, had a crazier summer schedule than ours. They went MIA. A couple of weeks later, my cousin Tony arrived with his family from Portland, Oregon. They would stay the night with us, and then we would all head down to Stonington for our annual August family gathering. Adding to the excitement: my ancestral home in Stonington had just gone into contract. My sister and I had one month to pack up the house that had been built by my father's parents one hundred years before, and that she and I had inherited from our father when he died of cancer in 1993.

I reached the logical conclusion that we should bring the chicks to Stonington with us. Why split up the family we had just created? We had already discussed having road chicks. Chickens do travel. People have show birds. And we should get them used to moving about, and us to moving them. Images arose of our menagerie—six chickens, our cats, Nick and Nora, and ourselves—all stuffed into our Prius, even as Carl's reaction to the idea cooled to arctic chill levels. He pointed out that we had no coop in Stonington. I countered pointedly that we had no coop anywhere, and berated myself for being so irresponsible as to have adopted these little creatures only to desert them. At Carl's insistence, I formulated and repeated a new mantra: *I will not let chickens rule my life. I will not let chickens rule my life.*

Fortuitously, Katie and David, our Angell Street house sitters, were available to house, cat, and chicken sit during our absence. As fortuitous, Tony had chickens in Portland. He knew how to keep chickens alive. With his help and by day's end, the coop and run were finished and raccoon-proofed, the chickens were exploring their new home, and I was doing a victory dance while scrubbing out the bathroom. Peyton, who was Best Friends Forever with Tony's eldest daughter Lulu and therefore visiting, thought I was totally awesome to give up a bathroom to chickens, but I knew there were deeper issues to be addressed.

For example: Just how powerful a role would I allow those prehistoric creatures to play in my life? They were not—disappointing but true—an excuse to stay home. They were not a fiscally responsible choice. They were, on the contrary, eating a lot of expensive food and still three months away from being old enough to lay a happy egg. If they lived so long.

We headed to Stonington for the family reunion, after which my sister and I, with Carl's and myriad friends' support, packed and divvied up the family heirlooms, furniture, and accoutrements that had filled the house that had been home to four generations of my paternal family, and it's amazing how much meaning plastic coat hangers can take on when letting go of something so much bigger and far less tangible.

On weekends, Carl would leave to jet-set off to a Tilt-a-Whirl gig.

Arizona. Vancouver. Winnipeg. Florida. By some miracle of love, we siblings were still talking to each other whenever he returned and at the end
of August, when two sets of movers arrived, one for my sister and one for
me. We might have maxed out our stress levels but we had succeeded in
moving out of that indelible house in three weeks.

After furnishing the New Hampshire house with all things Stonington,
Carl and I returned to Rhode Island. Thanks to Katie and David, we
found the pullets alive and well, though Rhoda Red's wattles had begun
to look more distinct than any shy and retiring girl's wattles should look.
And her tail a bit plume-y. Her crown, we noted, had always been distinct.
Her feathers shiny. I sent in a photo to MyPetChicken, asking if she was a
girl or a boy. MyPetChicken all too promptly reimbursed me for the cost
of the cockerel that was supposed to have been a pullet. Which was not
the point. I was worried about splitting up the family. Upon this *supposed*
confirmation of Rhoda Red, in fact, being Oh-Boy-Our-Boy-Big Red,
we debated: should we would move the six chickens to New Hampshire,
or give them all up?

Too black-and-white? But cocks, euphemistically known as roosters,
are not allowed in Providence. Even if we were on the city's outskirts, we
had neighbors who might be bothered by a cock-a-doodle-do or two at
three in the morning.

I wondered: Was the universe trying to tell me something? Because our
timber frame *weekend* house had, by now, been added to my list of worries.
Carl had asked me, and I had asked him: Does a newly built off-grid house
require oversight initially? Just for those first weeks, to be sure everything
is running smoothly? In short, Carl and I had been discussing going up to
Darwin's View for a longer period of time than overnight. Three, maybe
four months. The winter months. To avoid frozen pipes.

When I mentioned our quandary to my cousin Lise, she asked why we
would move to Darwin's View to live for the winter months. That was
ridiculous. We would, of course, shut down what needed to be shut down
of the new place and stay in Rhode Island, allowing Carl to be on the road
in peace and me to attempt to find what sense there was left in me. She said
I should wring Big Red's neck.

That was, of course, an option. Far be it from me to say never. But it will be a cold day in hell that I wring my chicken's neck, or allow anyone else to do so.

That being then. This being now.

Then, I saw no in-between. I felt sick at the thought of taking Big Red away from his familiars, thereby leaving Lola and CooLots to fend for themselves, without their go-to guy. Yes, yes. It's nature and natural, that pecking order. *Chicken rules*. And frankly, Lola was feisty enough to go up against Chipper and Panda any day of the week. And I had noticed that CooLots, even with her ridiculous feathered feet, had taken to challenging Ping, going chest to chest, feathers aflutter. What I didn't know was if I had the heart to separate them.

Or to separate from them. The girls and I had bonded. I would spend time out in the garden watching them show off their scratching and pecking abilities. They would help me weed and stand by, preening and gossiping, while I cleaned out their coop. Two or three weeks of homemaking and the girls and I had created a special bond. I never wanted to leave them.

But the phone rang, as it sometimes does. My uncle had died of a heart attack. We returned to Stonington for the funeral, having set up *ad hoc* chicken and cat care. Odd. We had no home in Stonington anymore. An adjustment. My mother, Xanda, Carl, and I stayed in a hotel for that long weekend. When Carl and I returned home again, the cats yawned and rolled on their backs to have their bellies scratched. The chickens, however, acted as if they had missed me. They were remarkably interested in me. What might I have to offer? Treats? Did I have any treats? Our girls had not yet matured to egg-laying age, and so I boiled them a couple of farmer's market I-hope-they-were-happy eggs. Hard-boiled egg is a chicken's equivalent of baby food. The result? *They loved me*. Whenever I stepped outside, the girls came running. Have you any idea how satisfactory that is, after years of the cats barely lifting a whisker in recognition of my return? Twelve bright and beady eyes uplifted. Heads tilted flirtatiously. *What 'cha got?*

What's a sentimental gal to do? Call me dramatic and ridiculous for anthropomorphizing six chickens when there are hundreds of thousands, no, billions, of chickens in the world, bred to grow really big, really fast, living in cramped, filthy, unnatural conditions until gathered up by a chicken vacuum and put into crates and ported to the slaughter house, where they are defeathered and decapitated (often in that order) at the rate of 140 chickens per minute. What difference would these six make?

All the difference. Because I could save them. I could keep them safe. That, at least, was in my control.

Thus proving the power of delusion.

Winter approached. Our brand new solar electric energy system in New Hampshire went into permanent default—also known as not functioning—twice. Big Red, our adolescent outlaw, was crowing with cocky pride and gusto. It was time. We moved—*temporarily*—to Darwin's View on November 27, 2012. Carl drove the brimming-full rental truck, and I drove our Prius, white-knuckling the steering wheel in preparation for the drive, near dusk, through snow flurries. Nick and Nora, in their cat totes, slept, indifferent to the occasional stirring of the chickens as they rustled contentedly in their hay-filled, extra-large dog cage. As we pulled out of the driveway in Providence, I noted the significance of the occasion: after years of talking about it, we had finally achieved what we had set out to do. We would have an off-grid home in New Hampshire.

Note to self: be careful what you ask for.

Our view of the Wapack Range at sunrise

II

2013 ADAPTATION

My office looks out to the Wapack Range. Every morning, as the cock's crow heralds the dawn, I have studied the peace of black and cobalt blue. And then a hesitancy as the sky and mountains become gradually more defined by the brightening horizon. A mix of yellows, oranges, and reds, and the black blue becomes violet and azure, gradually more intense—an impressionist's palette—until the sun peeks over the ridge, blinding and triumphant.

The difference on a cloudy, rainy, or snowy day? Gradations of gray that are all the more heartbreaking because they expose the fact that there is no black or white in life. No definitive answers. Not even death is final, except in our limited, human definition of it.

OFF-GRID DEFINED

The Jaffrey building inspector had said, early on in our planning process, that we couldn't do it, a rather off-the-cuff judgment that didn't sit right given the circumstances: a Jaffrey town Planning board meeting. We had approached that board to ask for a special exception, also known as permission, to put up a yurt. When we entered the room, Carl had a moment's hesitation. I looked at him. He smiled uneasily, nodded to Mr. Grodin, whoever he was, and took a seat. Mr. Grodin narrowed his eyes at Carl.

"Carl Querfurth?" he said.

Carl flushed and nodded. "Yes, sir. Hello, Mr. Grodin."

I learned later that Mr. Grodin had been Carl's advisor in high school. The one who would catch young Carl smoking in class, holding his cigarette out the window. Who, on more than one occasion, had stern talks with young Carl about rules and why they exist and why one should try not to break them. And here we were, asking to be exempted from one.

We explained to the board that we hoped some day to build an off-grid house, but, in the meantime, we wanted to put up a yurt to get the feel of the place. The board was skeptical at best, stymied at worst. A lot of people put up tents and then live in them for decades. How did they know we wouldn't do the same?

"I have no intention of living in a yurt," I, in my city business slacks and high heels, assured them. "This is temporary."

They narrowed their eyes, debated, and eventually looked to Dave Baron, the building inspector.

"They can put up their tent," he said, arms crossed as he leaned back in his chair, his square face set in a glum scowl. "But it has to be connected to a septic system and meet snow and wind loads."

Everyone sat silent.

"Does it say anything in the code about tents?" the board chair asked.

"Nope," the inspector replied.

"So if it's not in the code book . . .?"

"If it's not in the code book," the inspector said, "you can't do it."

66

"So the answer is no?"

"Yes. The answer is no."

We donned our coats, baffled and angry. Carl had spent a lot of time and energy figuring out the platform on which to put the yurt, and confirming that it met snow and wind loads, but to attach it to a septic system seemed counterintuitive. This was a tent. Carl brushed past the inspector without a word. As I followed him, Dave Baron looked at me and said, "You can't do it. Off-grid isn't possible."

"Yes, it is," I said, a bit sharply because I might not have been able to control the hot flush that coursed through my body and up into my cheeks. Nor could I prevent his obstruction of our every idea, not least putting up a tent on our property, but I wasn't going to let him think for one nanosecond that off-grid wasn't possible. "There's a house in Peterborough. We visited it just last week. It's off-grid. It's totally possible."

Now, then. Anyone who knows anything about the history of Peterborough and Jaffrey knows I had just raised the ante. Had I forgotten the golden rule? Never, ever compare Jaffrey to Peterborough. It upsets people. A decades-long and edgy competition exists between the two towns. Carl remembers as a youth biking home from a Peterborough/Jaffrey game and the bus of Jaffrey kids driving by. They spit on him—which proves Carl's forgiving nature. He has gone so far as to move to Jaffrey. Then, however, as I walked out of the town office building, I could only hope Dave Baron would consider it a challenge: if Peterborough could do off-grid, Jaffrey would do it better.

The option being, not at all.

All this to say that, when we arrived at the top of our hill in Jaffrey, with our two cats and six chickens, to settle in just for the winter, I sent thanks to the building code and Dave Baron because it was stark, and cold, and very, very windy. Yurt living? I wouldn't have done it.

Certainly, I would have vetoed the inclusion of the cats and chickens in that experiment.

I hope I would have.

That first week at Darwin's View, our attempt to create a home had a one-step-forward, two-steps-back feel to it, likely as not due to the site work being done outside; a case of a contractor's "two weeks" pushing a September project into late November. With all the disorder and noise of the excavator at work, we had to be sure that Big Red and his hens didn't get lost or buried. Thus, we set them up next to the house in their tunnel chicken tractor. Made of white PVC tubing and chicken wire and covered with a brown tarp, the chicken tractor kept Big Red and the girls safe and in one place, even if they didn't understand that it was in their best interest. Panda clucked nervous reassurances, CooLots moaned her sky-is-falling worries, and the other girls fluffed and scratched about, on occasion raising their heads to look at their strange and moving surroundings. They looked at me querulously.

"It's breezy," they seemed to say. "Is there no way to stop our walls and ceiling from flapping so alarmingly?"

I looked for a rock that wasn't frozen to the ground to hold down the tarp, leaving the girls to hop up into the nearly finished coop with Carl. He crouched in the plywood box that measured seven-by-five feet, three feet high at one end and two feet at the other. Screwing in the nesting boxes and the roost, he had Ping at his elbow, watching and advising. Chipper and CooLots judged the spacing from a distance, while Panda and Lola hopped onto the roost to be sure it was positioned at a suitable height. Much negotiation back and forth with CooLots "tsk"-ing and Panda and Lola cackling and scratching. Carl sat back on his heels, removed the screws clamped in his mouth, and looked at me.

"Could you please keep them occupied? I'm trying to get something done here."

Just in time, a roar and rumble from that large, yellow dinosaur in the yard, so naked of feathers, and emitting stench and noise. Big Red crowed his challenge. The girls hopped out of the coop and ran the length of the run to watch this new and exciting life event: the excavator. Whereas the chickens were currently on Tier One, near the house, the excavator was on Tier Two, thirty feet from the house and lower down by four feet. It sat in one place, but its arm would swing alarmingly close to the chickens' run, pushing piles of dirt toward them and then taking it away. Or it would pluck a huge rock out of the ground and lift it up, swooping it past as easily as a balloon. Wide-eyed, the chickens and I watched as the excavator leveled dirt, pushed boulders, and built stone walls, and it is fascinating to

Excavating near the chickens

watch the power of such a massive machine. The girls and I were transfixed as it accomplished what would have taken four bare hands months to do.

Big Red, as young boys will, scoffed. He challenged that yellow monster, especially when the arm of the excavator swung around and his beak was not a foot from the shovel that was five times his size and made of metal, not flesh and blood. Big Red, undaunted, scolded the shovel, marching past it and back again, pacing his line in the sand. He glared at me, making it clear he did not approve. This living situation was insecure at best, he crowed. I assured him that soon their home would expand to include a cozy coop with nesting boxes. I pointed, again, to Carl, who, in jeans and a work jacket and with frozen hands and runny nose, was hard at work, crouched in the coop that was built for chickens, not humans; this sizing issue was something we would come to consider over the years, but for now?

"Look!" I said. "Carl is tweaking your new home. It's almost done!"

While Carl finished up the coop outside in the frigid cold, I organized and cleaned the inside of the nice, warm house. By week's end, we had things relatively under control. That Saturday morning, there was no giant Tonka toy outside to distract us. The coop was connected to the run and covered with hay bales, and had been duly dubbed the Hay Chalet. I made our morning *cafe lattes* while Carl fried up four eggs (still not our happy eggs, but we were getting closer to that day), toasted two slices of bread, and blended up a fruit shake. Over breakfast, we assessed what a huge amount of energy we had used to make breakfast: Espresso machine.

Toaster oven. Blender. A perfect storm that had caused the generator to start up. How joyous, to have that backup, and yet weren't we supposed to stay within the limits of our solar-powered batteries?

We contemplated that dilemma on that gray morning. Mount Monadnock lay hidden in a white and gray veil, a snow shower that we watched approach. Nora curled herself into a smaller-than-she cardboard box. Nick sprawled on the couch. A fire blazed in the wood-burning stove, warming the fifteen hundred square feet of the house to a balmy seventy-five degrees to counter the outside cold. Carl noted we'd have to experiment, balance the tightness of the house envelope with the munificence of the stove. In the meantime, dressed in T-shirts and pajama bottoms, we breathed in the perfection of that off-the-grid moment.

"Off-grid," simply put, means not connected to the national power grid. But it also suggests a variety of lifestyles. For some people, it signifies a pioneer-ism and love of dirt and plants and connectedness. Who needs electricity or showers? There is the sun. There the rivers and streams. To others, off-grid might suggest outhouses, one-room hovels in need of vacuuming and a twelve-volt battery to run the single-burner electric hot plate. Hunting and gathering, or perhaps veganism? Still others think of preppers, people who are prepared for the apocalypse and/or the demise of society as it is currently known. They have stockpiled food and ammunition and, hopefully, have remembered long-forgotten skills that will be needed in the coming debacle—or, at least, a can opener. To yet others, like our building inspector, it means a yurt unconnected to a septic system, with insufficient support for snow and wind. Myriad images of hair-shirt off-grid, none of them being ours.

We are twenty-first century–American off-grid. At the time, Carl's and my house looked like a normal house. At fifteen hundred square feet, it had four sides, light switches, 0.8-gallon flush toilets, and a washing machine, though no dryer. Wi-Fi internet service and hot water showers were the norm, and our woodstove had a backup consisting of radiant-heated floors, a form of heating that we avoided because it ran off our propane gas, as did our third backup power source, an 8.5-kilowatt generator. Redundancy is good when living off-grid.

Our generator—upon which we depended on sunless days because we have no wires to connect us to anything beyond our little hill—went straight-up splat against my image of off-grid. To me, off-grid meant fossil-fuel free. Propane gas brought to my mind images of once pristine lands raped of their flora and fauna, blackened by tar, and cluttered with pipelines, ravages of nature as heinous to me as vegetarianism is to some beef-eating Americans. That image was one of the reasons we moved off-grid. We didn't want to be part of that system.

Eh, voila! Once again, my hypocrisy rose like a phoenix. Carl and I might have appeared to have lowered our carbon footprint by being off-grid, we might have deluded ourselves into thinking we were taking ourselves out of the fossil-fueled system, but an analysis of our externalities proved that, despite all our efforts, Carl and I epitomized our society's lack of sustainability and love of convenience. We maintained two houses, one of which might have been off-grid, but what of the devastation wrought by its building process? Hundreds of trees timbered. Soils uprooted, deeply disturbed, and, in places, eroding. Solar panels that are made of rare and precious metals, and we couldn't ignore the gasoline used to get us between our two houses, to gigs, to the grocery store, and to other points of interest. Who was fooling whom? The only real difference between us and an on-grid house was that, when there were blackouts or power outages elsewhere, our lights still shone.

Usually.

Few trees had escaped the 2006–2007 timbering of our hilltop. Those that remained were debilitated by the ice storm of 2008. Of those that survived, we had some removed to provide full sun exposure for the solar panels. Thus, there wasn't so much as a blade of grass to stop the voluminous breezes coming over Mount Monadnock from slamming against the house and battering the chickens' Hay Chalet. Every night, three to five times a night, I would wake, my body tensed against the seemingly mortal danger of the winds that hammered the house. They convinced me that we were fools to be there.

Early on in our just-for-the-winter venture, while attempting dinner conversation over the outside din, I commented that the house sounded remarkably like an old, wooden boat, one of those Spanish Armada galleons creaking and moaning in the middle of a typhoon.

"We aren't going to sink," Carl reassured me.

"Obviously," I said. "But the house might get blown down."

"Nothing is going to take out this house."

"That's what they said about the Titanic," I remarked, just as the French doors that faced Mount Monadnock bowed in under the force of the most recent west wind blast and we heard a loud pop.

"Just the house timbers settling in," Carl said as we both sat back down in our chairs and another gust hit the house. I found it a pyrrhic victory that Carl finally looked nervous.

Not that he admitted it. A very New Hampshire trait, to be unconvinced by weather.

"Ah, yuh," you'll hear when you comment on the blizzard conditions, the ice, the negative wind shear. "That's what you might expect, it being wintah."

And so at dinner parties, when guests comment on the wind, the doors bowing, the good fortune of an overbuilt house that could never be taken out by mere cyclones, I have learned to say, "Ah, yuh. We've had worse."

At which point people ask why we don't have windmills.

"The wind's dirty," we reply in unison. Coming over the mountain and swooping down and up, the wind is blustery and uneven. It might do to a very expensive windmill what it did to our anemometer the first day we put it up. Carl and I had sipped our morning elixir of Santo Domingo espresso roast *cafe lattes,* and watched, rapt, the anemometer's receptor screen. It showed the wind averaging thirty to forty miles an hour, forty-five, and then a gust slammed against the house. The screen flat-lined. Very bad timing. No anemometer and Blizzard Nemo tracking its way toward us.

That first winter at Darwin's View, the winter of 2012–13, had more BIGGEST STORM OF THE CENTURY! alerts than had ever occurred in the brief history of humankind and the Weather Channel. Somewhere in our country, someone was getting slammed. More often than not, it was us. Having spent years worrying about the world's lack of snow due to the warming climate, it felt reassuring to be in the thick of it again.

The beauty of snow? You can see the wind that is so eminently present at Darwin's View. Shrieking ghosts. Pale viragos chasing scampering, diapered cherubs across the fields. Whirlwinds bashing against tumbleweeds that suddenly calm to nothing, only to rise up as fields of whistling grasses, oceans of frost. Whorls of energy that left parts of our hilltop (the chicken coop) six feet deep in snow and parts (the cars) swept free. And still the wind. Still the snow. On top of a hill. Off-grid. With solar panels, not windmills.

Winter days are short, the result being less sun. And though daylight might come, when it's snowing, there is even less less-sun. Like none. If you depend on solar panels for your electricity and there's no sun, the batteries that would usually run the electricity in the house run down. Fortunately, we had our propane generator backup system. Every third or fourth day of no sun, the roar of the generator, typically starting up just as I turned on the espresso machine, assured us that all was in working order. Until, one day, it didn't. Not that we noticed, because the batteries still worked. It's just they weren't supposed to get used down to the very last drop. It's not good for them.

One of the questions we were asked by our off-grid energy consultants in the planning stages was this: When you arrive at your off-grid house, do you want to flick a switch and have the lights turn on, or are you okay with having to check how the system is working? My response: flick a switch. Carl's: How does it work?

Carl's and my yin/yang approach: whereas I prefer to spend my time on the whys of life, he asks the hows, which is far more helpful and relevant when in the middle of a blizzard and the generator is not working, a fact that Carl finally noticed when he went to check the batteries in the basement and noted that the inverter had gone into default. Thus, while I remained inside, worrying that Carl would get lost in the woods to our north, to be found in the spring inside the trunk of a dead tree where he had sought protection, Carl opted to take some initiative and *a posteriori* test the wind velocity by troubleshooting the generator's untimely somnolence.

I didn't mean to compare. Comparing is unhealthy. It causes resentment and anger. But there we were, two months into our little experiment, all by ourselves on our lovely hill with all kinds of amusements, like wind, snow, and ice, and I had the stark realization that I preferred city living. Where we didn't have to stoke a wood-burning stove five times a night. Where we didn't have to drive for fifteen minutes to arrive somewhere, *anywhere*. Where the walk to a market is not eight miles and past a shooting range and, thus, where a black- and brown-toned city wardrobe doesn't have to be replaced with florescent orange. Most important of all, though, in a city, I could delude myself into thinking I could take care of myself. I could pretend, as does every good American, independence from things I am entirely dependent on.

But *in the country,* I am surrounded by acres of land we were supposed to be *stewarding* and living with a man who thrives on being outdoors in frigid cold, proof being his conviction that skiing in a blizzard is the penultimate form of fun. *In the country,* there is too much to learn and know, none of which includes sitting down in a cozy chair and reading a book. Thus, the dis-ease I had felt that first time Carl brought me to New Hampshire had not changed. *Au contraire.*

And so I compared. For just a moment, staring out into the darkening skies, deserted by Carl, who had already either died of a turbo-charged electric shock from the generator or disappeared into a Darwinian Bermuda Triangle, thereby leaving me to fend for myself, I compared and felt resentful. I had given up too much in this move to New Hampshire. I was unprepared. *It wasn't fair*!

And then the roar of an engine and my resentment melted into gratitude. Thank goodness for our Subaru. Carl was warming it up to drive me back to Rhode Island, right? No.

Bought in 1998, the Subaru was a tank and a truck all in one and, by 2013, in its dotage. One symptom: It would lock, then unlock its doors. Lock, unlock. Lock, unlock, lock, unlocklockunlock. At the same time, its lights blinked. Blink-blink. On, off, on, off, onoffonoff. All this while the engine was off. Thus, the battery would die. Time and again, and this time, too. When Carl stepped off the porch to go to the generator, the Subaru, perhaps out of joy, like a lovesick puppy, began to lockunlocklockunlock and blinkblinkblink, then died. I would have left it to sit. There was a blizzard going on, after all. Carl, ever the hands-on guy and fully energized by the weather, jump-started it (thus the roar of the engine of the car, not the generator) and left it to run while he dug the generator out of the snow. At which point, he realized the Subaru (lockunlocklock) had locked him out.

While he attempted to jimmy the lock, ice forming on his long eyebrows, I braved the elements to shovel the girls out of the Hay Chalet for the third time that day. Upon arriving at the door, I dipped into the run. Hay bale walls blocked the wind, leaving me to wonder if we shouldn't

74

Our first storms, from outside and inside the coop

have used the same for our house. Also, the hay that served as the floor had begun the composting process. My unwitting success at deep-litter mulching made for radiant-heated floors. The hens cooed and clucked. The atmosphere was gentle and soothing. Reassuring in their visceral calm, the chickens were *being*. No worries. Their only concern: Where were the treats? At which point, a peck, a squawk, and feathers flying. Big Red chastised and directed until the hens settled back down around him, and me too.

Kneeling down in the run, I listened to the hens rustling and studied their social etiquette. Panda and Lola groomed themselves on either side of Big Red. CooLots uh-ohed off in a corner. Ping and Chipper, side by side, caballed. Death felt very far away. The only suggestion of necrosis was Big Red's comb, which was blackened by frostbite. For all the time I had spent googling possible remedies, the fault pointed back to the human element, drafts, and moisture. Big Red looked at me with disapproval. What good was I if I couldn't heal his comb?

Bowed, I crawled back out of the coop and faced the buffeting, shrill winds of Blizzard Nemo. The ice cold reddened my nose and pinched my cheeks. Struggling to maintain my footing against the wind, I watched the snow sweep across the fields. It was a stunningly beautiful winter afternoon and my self expanded in the maelstrom, alive. Noting the car's headlights in the darkening skies, the running engine, I stomped the snow off my feet and headed inside. Carl hovered by the stove, warming his hands. In the dark. I reached to flick a switch.

75

"Don't do that," he said. "The generator isn't turning on."

"Excuse me?" I asked. Though I knew. Off-grid. No sun. No generator? But I am the one who, from day one, said I wanted those light switches to work because I live in the twenty-first century. Ever with my priorities in order, I asked, "Will we be able to use the espresso machine tomorrow morning?"

Carl called the generator guy, who, in the course of his interrogation of us, asked why we hadn't built a box around the generator to prevent it from sucking snow into itself, thereby freezing the system. And was the propane gas tank shoveled out? It needed air, too.

Now then, *who would have thunk to build a box around either of those objects?* Not us, in part because the generator company that installed the generator had said nothing about building a box around a generator that already has a metal box around it, one bought specifically for its added protection. Putting a ventilated box over a ventilated box? That's like having a piece of equipment that tests wind velocity burn out on the first slightly windy day it's up and running.

Thus, we had no anemometer for the two worst snowstorms in history.

Thus, we had frozen our generator, and the emergency generator company's on-call guy didn't have four-wheel drive.

If one were to be judgmental, one might ask why an emergency generator company based in New Hampshire doesn't provide its employees with some semblance of four-wheel drive transport. But we were not ones to judge. At least, we tried not to, so full of faults ourselves. Besides, this was an adventure! We were adapting! Evolving!

That weekend, we did no laundry, nor did we vacuum. We allowed ourselves one *latte* per day and didn't charge our computers or cell phones. We had candlelit dinners with no technology to distract us. We couldn't remember the last time we'd had a candlelight dinner.

Ever the "how it works" guy, and using a solar flashlight, Carl went down to the basement to see how many watts we were using: 150. The refrigerator was the only thing electric running in the house. We had only candlelight and the light from the wood-burning stove to see by. It was very quiet. Still. Because this was life off-grid in the twenty-first century. All the luxuries of a regular house, when the sun shines.

The sun not shining in winter

ASCENT

Mornings, I woke up to his crowing. "LET ME OUT!" "WAKE UP!" "COME ON! LET'S GO!" "DANGER! DANGER!" "DON'T EVEN THINK ABOUT IT! GET UP!" "YOU LOOKING AT ME, KID?"

Big Red had adjusted nicely to his environs. He had his coop and his hens and a natural enemy, the wind. It disheveled everyone and he faced it down whenever possible. Like a matador with his cape, Big Red would turn his shoulder toward the daily outrage. Leaning in against it, he would thrust his neck forward and crow mightily at the powers directing their forces toward him. Pre-sunrise to sunset, we could hear his threats and promises. He didn't bother with niceties. His hens needed protecting.

Big Red

Of course, the hens and I had no delusions about who was in charge. It was unspoken and understood, as are so many power struggles: Big Red would eye the sky, the fields, and the humans around him. He would strut and dance, cluck excitedly at a tasty treat until Ping, CooLots, Lola, and Chipper would come running, surrounding him, to his great

Panda and Big Red

satisfaction. Until Panda would gently cluck: time to retire to the run for a dust bath and nap. The girls, their feathery bottoms waving in the wind, would follow her to their beauty rest, leaving Big Red to pace and grumble in solitude, daring the next gust to blow.

We humans, in the meantime, prepared for my mother's winter visit up to Mud Mountain, as she dubiously called it. The unusually sharp and steep curve of the driveway about two-thirds of the way up terrified her. I was okay with it so long as I didn't have to drive. I could pretend bravery, my eyes shut, as we rounded that corner, my mother clutching the seat and crying out, "Carl, be careful! We're going over the edge!"

My perfect mother. Plus age. Add Parkinson's disease. Whereas for decades she was hard to catch, in New York City, traveling in Europe, ever busy, now she had time. Parkinson's gave my mother back to me. Granted, she couldn't walk through the woods, her face flushed with cold and the vigor of life as it used to be on our wooded walks when I was a girl, but her presence soothed me. My shoulders would relax from their tensed position as I soaked in her presence, reached for connection. Words and ideas had always served that purpose with us. In *tête-à-têtes*, mother to daughter, we talked about everything from writing and work to family upsets. I fed on those times together and was always hungry for more.

Too bad. By the time Carl and I had built Darwin's View, the shift had begun from mother/daughter to daughter as caretaker. A natural event. People age. We slow down. But Parkinson's disease and its medications sped up that inevitability. My mother's peers still gallivanted about, traveling and working and conversing, animated, fascinated: exactly how my mother had always been. Until she wasn't.

She chose to take the medications, in hopes they would slow the disease. From the very first day, she was lost. Her desire to live and be active conflicted with the reality of a devastating mental fog and physical exhaustion. Her eyelids fluttered as she struggled to keep them open, to think clearly. Gradually, she slid into dependence. What began as a hiring of a personal assistant, who would juggle doctors' appointments, studio visits, physical therapy, and medication ordering, evolved into overnight care and more hands-on assistance during the day. By degrees, she gave up control and then regretted having given it up, as if it had been an actual choice. It wasn't. How often was I frustrated, on the other end of the phone, talking her through pill time, trying to convince her that she had to take the medications, that, no, she had not yet taken.

"Please take the pink oblong one. Is it in your mouth?" I would ask because I couldn't see. She was of the generation that mistakenly duplicates hard drives instead of hitting the Skype icon. "Now drink some water." I would listen for the glug. "Did it go down?"

"Of course it didn't," she would snip, irritated by the stupidity of the question. "I'm not taking it."

A half hour, an hour, on the phone to support and guide her. Four hours later, again. For years and years, she trooped on. Was it the disease or the medications that she fought? Medicine as practice and our mother as guinea pig, we would never know.

Funny how that happens. We make choices that make perfect sense at the time—cars, not horses; plastics, not cloth or glass; fossil fuels, not renewables—and onward we go. Things glide along until they don't, at which point, too late, we look about us. A lunge for the past . . . and then we trudge forward, dealing with the consequences within our limited capacity.

Just so, my mother. Tallyho! She still had relative independence. She could travel, although no longer alone, and certainly not by train because the hustle and bustle of Penn Station evoked images of her lost and afraid, or pushed in front of an oncoming train. And so, if Xanda or I were unable to drive her ourselves, we would hire a driver—who used to drive our mother's mother—to take her back and forth, Stonington to New York

Mom with Big Red and Ping

City. And now to New Hampshire. A long drive on a good day, my mother arrived tired but game for a country-life experience.

The sky was crystal-clear blue. Crisp winter air. Snow on the ground but little ice. Carl and I bundled her up in a forest-green duffle coat, a sturdy pair of winter boots, mittens, a black-and-white striped hat and a purple scarf. We led her out to the great outdoors and the Hay Chalet. With Carl and me hovering, she walked without assistance to a frozen hay bale outside of the chickens' quarters and sat down. I handed her a handful of kale. To her horror, the girls swarmed her. Worse, so did the terrifying Big Red.

"God help me," she said as the chickens pulled and ripped apart the kale, ravenous for greens in winter. Mom held the leaves tight in her grasp and laughed nervously. "He's not cock-a-doodle-doing."

Big Red chortled and tidbitted.

"He's sort of cock-a-doodling," Carl said as I held up a perfect egg and chortled, too.

"I'm just a girl," Mom told Big Red, who danced toward her, his form of flirtation. Mom wasn't buying it and looked away and off into the distance, as if looking could transport her to another, less invigorating place. Like a New York City subway stop, where strangers pushed and shoved,

80

begged for change, or shouted out prayers, threats. That being far preferable and familiar to Mom than Big Red, who, with Lola staunchly by his side, crowed to announce the end of another day. The girls retreated into the coop. I held out an arm to Mom, assisting her as she tiptoed over the snow and back into the house for dinner.

Too brief a stay, two weeks. I returned to New York with Mom to help settle her in. To attend doctors' appointments. And then to Rhode Island. And back to Darwin's View.

Big Red and Lola

Tim stopped in to check on the house and scold us for not using the ERV. The Energy Recovery Ventilation system was installed to exchange used air inside with fresh air from the outside.

"How can you tell we aren't using it?" I asked with a "told ya so" glance at Carl, who had copped an attitude against the ERV because it was, in his opinion, a waste of energy. Tim, however, responded that our floors were beginning to buckle, ever so slightly. The windows were sweating. Tim explained, again, that the house is tight and the timbers fresh, and so there was a lot of moisture that had nowhere to go. And no new air could get in. The result? Once we use that air with activities like breathing, it's gone. He told us the story of a cat trapped in an Igloo icebox. The story did not end well for the curious feline.

"If you don't start using the ERV, you might be the cat," Tim concluded.

Carl and I are experiential learners. A few days later, I was busy doing my own form of nesting, which meant moving around my office in hopes of enlarging it beyond the square footage that it was; my brain had no room to expand. It being a bright-blue-sky sunny day, Carl decided it was time to equalize the battery bank.

Because we are off-grid, our solar panels charge our batteries that run the house. Lead acid batteries, which are what our system had at that time, need to be balanced every few months because each of the batteries—like people—is unique. Some work harder than others. With time, the batteries become unbalanced. To correct this, they need to be overcharged. Overcharging burns off all the gunk that builds up, and the batteries are left "equal." I left the details to Carl, that being his job and interest. Remember? I flick the switches. He checks and verifies that they work.

Up and down the stairs he went, like a kid checking his science experiment, until the batteries were made equal. At which point, tired out by his own version of StairMaster, he yawned and lay down for a nap on the couch. It was a sleepy kind of day, the sky already preparing for dusk at three in the afternoon. Naps are good in that kind of weather. Hibernation is what winter is made for. Nick reclined on the floor next to the woodstove. Nora snored on my desk. The book I read was a bit dull. Drowsy. Heavy head. Whole body relaxation. Wouldn't a nap be nice? I lay my head down on my desk, figuring a quick snooze would feel really good. Pretend I'm a cat. Contented sigh.

RAARAARAARAARAA!

Alarms! Smoke detectors! Nick scrambled and skidded across the wood floor. Nora knocked over a water glass on my desk. Both tails, fully fluffed, disappeared under our bed. Carl jolted awake and looked about him, befuddled by the noise. Once I had recovered from my own near heart attack, I glared up at the shrieking alarms. No, Carl and I noted. Not the fire alarms. The carbon monoxide alarms.

Thirty degrees outside? We opened the windows and doors of our oh-so-airtight house to let in air. Counterintuitive, perhaps, to leave a window open, thereby defeating the purpose of an airtight house? Then turn on the ERV, thereby using energy at a time of year when the sun has gone on vacation, leaving us to live in a damp, white sock for days on end.

That winter, Darwin's View swallowed me whole. It surrounded me, a stark expanse of land. Staring out of my office window at all the site work that had yet to be done, with Carl in the basement practicing his trombone and puttering contentedly about—checking the batteries, repurposing the leftover trim wood to make squatty potties and cheese cutting boards, reading his *Home Power* and *Biocycle* magazines—I felt lonely in a breathless, existential way. The wind buffeted the house relentlessly.

The Girls trying to get inside

Snow dervishes danced across the field, and I was eye-to-eye with Big Red on the covered porch. With the girls settled around him and staring into the office, he would cock-a-doodle-do. I would ask, *Can't we go home to Rhode Island?*

Big Red was a vocal cock. He couldn't go back. And his compatriots had begun to lay their happy eggs. How could I leave them? How could I stay?

And more snow. We shoveled out the generator, fuel pump, cars, and chicken coop. The wind moaned. So did I. The oversized woodstove kept the house toasty warm but I did not feel safe and cozy. On the contrary, I had the visceral sensation that the house was, in fact, moving. I imagined hay bales tumbling across the field like tumbleweeds, and chickens, too. I lived to the tune of a high-pitched scream that sounded remarkably like an overtone of the howling wind. I continued to wake with my body tensed in full-flight response. But where to go as the gales slammed up against the house from over the mountain? Direct hits. The house shuddered. One night, Carl woke me with a shout.

"The roof is peeling off!"

Orange and red shadows on the walls. Fire! We both jumped out of bed. The cats lifted their heads, irritated by the midnight stampede not of their own making.

Carl had had a nightmare. The reflection of the woodstove's flames on the walls had fed his confusion upon waking. Everything was fine, but it left me wondering whether he was, in fact, as comfortable with our house placement as he claimed. Did he, too, at times wonder if the gods were testing us to see how much we could stand? Or did he think, as he claimed he did, which was almost scarier to me, that the winds are arbitrary, without conscience, and uncontrollable? Doom thrummed in every buffet and howl. Nature used to be relatively predictable. Winter followed by spring. But change was coming. It was here. There. Everywhere. My neck and shoulders ached in anticipation. I read *War & Peace* along with really depressing books on the taut braiding of democracy and capitalism, the former choked by the latter because profit dominated nature and it wasn't cost effective to save the world.

Carl and I had played our part in that calculation. Mostly for the view, but in part for the money to pay for our 9.2-grade driveway, we had murdered countless trees when we first bought the land and implemented our forester's forest management plan. The top of our hill, in particular, had been cleared and the tippy-top five acres stumped. The work had been done while we were on a trip to California. When we returned to Jaffrey via Providence, the sight of the ravaged land was sobering. We hadn't understood how dramatic the culling would be. Months later, the true extent of the slaughter was brought home by a dowser.

David Yarrow was already at Darwin's View when we arrived, as ever, late. He had taken that time to go for a walk in what woods were left, and emerged, tall and thin, with a straggly white beard and a damning expression.

"I've been with the tree spirits," he announced. "They are not happy. Mother trees. . . ."

He stopped to breathe and look about him. At the view. The open field, the contours of the land. I got a sick feeling in the pit of my stomach. This was the first I'd heard of mother trees, the oldest and biggest trees in the forest. As I would learn, their roots are fungal networks, mycelium threads weaving a cellular path from tree to tree, plant to plant, vibrating energy through the earth. If we listened, we might hear it beneath us, in the soil, communicating, feeding, sharing information and nutrients throughout the plant and animal world. The world as a system in a greater system, the universe, and everything in it connected. The balance of nature working, even as *Homo s. sapiens* did seemingly everything we could to tip it.

Us, too. Carl and I had logged the property in what we thought was an environmentally sound way. Instead, we had caused the demise of three-hundred-, four-hundred-year-old trees. Our dowser knew because he had seen the trunks.

"Look!" he said, sweeping his arm to indicate our devastated hill. "Those whorls in the grass are the dragon's footprint. He is angry."

I looked about us, breathless. Dragon? I imagined Puff and his magic. Flames coming from his nostrils. His pterodactyl-like wings flapping . . . thus the wind! It was a dragon having a temper tantrum. I couldn't blame the dragon. Once a tree is down, you can't get it back. Human ravagement is forever and—.

"Could we discuss where we're going to put the well?" Carl asked, because he saw only where wind had flattened the grass.

David had a unique perspective. He had been electrocuted while moving a house as a young man. As a result, his hands were bent at uncomfortable angles, his body contorted, as if still frozen from that frizzling zap. His skin had the pale, smooth appearance of a healed burn wound, which it was. Looming tall, he carried with him a connection to other things. One wondered if he had been there and back, talking to the light at the end of the tunnel that those who have died and returned talk about.

According to astrological characteristics, air signs, a.k.a. Libras and Aquarians, tend to be unaware of the psychic energy around us. Certainly I didn't see energy swirling about in any set form—dragons, mother trees, or wood nymphs. But that idea comforted me. I might make friends with dragons and mother trees and wood nymphs. And even if I were oblivious to the swirling energy beyond, calling it gale-force winds or spring breezes, I was willing to listen to this guy's perspective on life because he was going to tell us where to put the well.

He stuck a wire coat hanger between his frozen fingers and proceeded to walk around, commenting and telling stories as he moved about the land.

"The Monadnock region is a significant place. There are a lot of water crossings here." He paused at what could arguably be called the top of our hill and looked about. "Lots of water is passing beneath us, strong currents. Ley lines abound here."

He continued to walk and explained that ley lines connect ancient monuments, from Stonehenge to the Great Pyramids to, yes, Mount Monadnock. Manmade or natural, they all hold spiritual and mythical energies that connect us to the earth and the universe. And you don't want to build on a crossing of water.

"A crossing here," he said, walking with his stick. "And here."

I picked up twigs and put them where he pointed. "It's bad to build where the waters cross as there's too much energy going about. No calm to settle in." I listened closely as he told a story about a woman who'd called him in to work with her. "She built her house over where waters crossed. I told her not to but she claimed that was the best site. Deep underground rivers crossed where she put her bedroom, exactly where her bed stood, her pillow. She developed a tumor in her head. Too bad she didn't listen to me. And here."

I placed another twig and noticed that in the wind, the twigs didn't have a lot of staying power. I moved on to small stones, a cairn of pebbles, desperately trying to keep track of all the water crossings. There and there. Here. But some of the spots where he pointed would be bad and some would be good. And there were a lot of stones on the property. Who knew which ones had been placed by me, and which by nature? What if we built in the wrong place?

"Could you just tell us a good place to drill?" Carl asked. "Specifics?"

Our dowser determined three or four locations that would be good. We asked him to pick the best. He dropped a small stick where, he said, we would hit water, one hundred and eighty feet down, flowing at forty gallons a minute. But it had to be there, not a foot to the left or right.

We marked the spot, and it was time to make amends to the tree spirits. A friend from Providence had arrived, as had the two people who had connected us to David, and our neighbor, Terry Landis. Our dowser sang to the tree spirits. He chanted age-old chants and was joined by our friends. I watched a bluebird surfing an updraft. The mountain seemed to have forgiven us for the moment, if only because it had forgotten us. The tree spirits were quiet. Calm. For that moment, peaceful. Priceless, beautiful peace, now that the loggers had gone. The dowser asked the dragon to forgive us and stop stomping around the property.

"Lunchtime," Carl announced, standing up and stretching. He is flexible when it comes to ideas, if not physically; once I convinced him to go to a gentle yoga class with me and he sprained his ankle. So sitting with legs crossed is not his favorite pastime, and, apparently, he could tolerate feng shui, bells, and sage smudging, but dragons? His New Hampshire, let's-be-real heritage had kicked in.

As for myself, I wasn't quite sure what to think. This being a special place made sense because it was, and our friends seemed to be fine with the differentness of the ceremony, and sitting in a circle with people who care enough about nature to be still and enjoy the moment felt holding, connecting. So what if our dowser believed in tree spirits and dragons? An awful lot of people in the world believe as passionately about other spirits and gods and mighty powers. Who was I to say any one of them was right or wrong or crazy? And who were they to say I was?

Carl and I might have been fools to build on top of that hill and not further to the east, back in the field, to allow for the wind force to die off, as every professional we asked had advised. But our intention in building on that hill had been so we could see the mountain.

And we might have been tree murderers for logging our land, thereby infuriating tree spirits and mountain dragons, but I would be the first to remind them, if ever I met them, that one hundred years ago this land was pastureland and orchards; that Carl and I intended to restore the soil and maintain the brush that was growing up, thereby providing rare habitat for turkeys, deer, and migrating birds; that we hoped above all to help, not to harm.

But the truth? Maybe there are no dragons or tree spirits or gods over-seeing things, no reason for evil and pain. Maybe one's parents are not perfect. Maybe humans aren't. Maybe we, as a species, have caused this planet's trauma and destruction. Maybe Americans, with all our potential, are not following in our founders' footsteps and doing whatever we can so that future generations might have some degree of joy in life, liberty, and the pursuit of happiness.

Hard to imagine some maybes, especially when incessantly moving about. Providence, New York City, Jaffrey, and around again as the un-pleasant consequences of my mother's Parkinson's progressed, a lowering baseline remarkably similar to that of Mother Nature. Words disappear-ing. Species dying. Less resilience. More susceptibility to confusion and chaos. I tried to ignore it, pretend it was normal. It was too scary not to. Besides, what could be done even if I looked reality straight in the eyes?

As I hammered a square—me—into a circle—Darwin's View—I sought a way to repurpose myself. Sick at heart with missing my home in Providence, I had to do something to make sense of that choice. Hacking and cutting away, I created a new story to tell myself: I was at Darwin's

View because the world was on fire. Politics and religion. Climate change. I saw the world's pain and wanted to stop it. I had to start somewhere. Mother Earth was at stake and the spectator seats were not comfortable, as close to sea level as they were. To me, there didn't seem to be an option. I had either to stand by and wait or to do something.

The problem was, everything I did seemed to contradict what I claimed I wanted to do. How to shift the world away from what I considered the evils of materialism, unsustainable excess, denial, and hypocrisy, all of which could be called the causes and symptoms of climate change, when I epitomized all the above? Two homes. Driving more than I had ever driven in my life. Drinking (fair trade!) coffee shipped from the Dominican Republic. Financial safety when most had none. I saw the myriad conflicts. Every day, I asked myself, Where do I begin? How line up my actions with my principles? How challenge myself to make the necessary changes and sacrifice when I didn't have to? At which thought I'd look out my office window to see a very alarmed CooLots running through what someday would be a garden as fast as her feathered claws could carry her, and a hawk swooping down on her.

"NO!" I shouted, standing up from my desk. CooLots, and the hawk disappeared around the corner. I ran outside. No hawk or chicken in sight. I looked in all the usual chicken hang-outs: the porch, under the cars, off in the fields. At last, in the coop, I found CooLots, huddled along with

one, two, three, four, five other chickens. All accounted for. A near miss, that clash of nature's natural selection with my protected world, and proof that chickens are an excellent form of distraction from uncomfortable questions.

CooLots

The Girls and me with our greens

WHAT I KNEW ABOUT GARDENING

I have always known, and am known for the fact, that I can pull together a dinner party for ten given an hour's notice. At least, I could in Providence. At Darwin's View, our refrigerator was too small to pack in such ample provisions for crowds. But we had pasta, rice, and potatoes sufficient to last us a week or two.

Greens were the problem. The chickens and I love our greens and felt the dearth of them by the tail-end of winter. I fantasized about a greenhouse. All I would have to do was walk out of the house and into the it-would-have-to-be-reconfigured-again chicken coop, through the as-yet uncreated, wind-protected run, to the not-yet-existent greenhouse, and there would be the girls, pecking and scratching the sprouts that I would rotationally, and with entirely un-me organization, plant for their benefit. I would pluck kale, lettuces, Swiss chard . . . and greet Bessie, the as-yet-unadopted cow. Her body heat would keep the chickens warm in the winter. Big Red and I wouldn't have to worry about his comb, which was getting progressively blacker, on some points white, with frostbite, and the temperatures were falling. I was eyeing the Bag Balm, and Big Red was eyeing me, clearly remembering our too intimate moments when I had applied that goo to his comb in the false hope it would soothe the discomfort of frostbite. Instead, the insult of hay stuck to his comb and flopped into his eyes.

This entire scenario confirmed that we wouldn't get a cow. If I got as anxiety-ridden over the chickens as I did, what would I do with a cow? And two pygmy goats. Piggies and a few sheep. Maybe a miniature pony, a donkey, alpacas. An elephant or two. Whales, unfortunately, were out of the question; not enough salt water or plankton.

I tried not to think too obsessively of all the animals, human and otherwise, who needed care-taking. I wanted to adopt and protect them all. That would necessitate feeding them. Grains. Vegetables. If one followed that line of thought to its inevitable conclusion, gardening, a form of farming, was in my future.

My dreams often dance independent of reality, perhaps a result of the attention deficit disorder that I had been diagnosed with in my twenties, but what poppycock! Mind over matter, I powered through, getting half the amount done in twice the time one would expect—and yet! Look at all the things I hadn't intended on doing that I got done, too. At night, though, I would mull and mope about what I didn't do that I said I would do. Example: create a garden worthy of Versailles. Instead, my inconsistent attention to detail left things more . . . natural. Tasks like keeping record of where I put what seed or plant. Watering. Weeding. These had always been shunted to the side in favor of writing, practicing flute, and the busyness of business, be it political, financial, familial, or social. But hope springs eternal! Mud season arrived, New Hampshire's "shoulder season" that typically shows up in April. We left the house and chickens under the care of Sarah and Peyton and returned to Rhode Island, Carl to demo the master bathroom that we had decided to update from its pre-retro 1950s plumbing, and me to set up that year's Beachmont Avenue garden, because what better way to fill my already filled time than gardening? It's so stress free!

In the basement, I surrounded myself with the variety of seed packets I had ordered, a dozen books on farming and agriculture, and the full squadron of my judgment fairies. They sat on my shoulder and whispered in my ear. *You're not a real gardener. A real gardener would have had her seeds out and organized weeks ago. She would have already started her onions and leeks, her broccolis, cabbages, and lettuces. A real gardener would not dread going outside and poking about in her garden beds in the raw and cloudy weather.*

I battled self-doubt, not least because a *real* gardener, a wanna-save-the-world kind of gardener wouldn't use Miracle-Gro, two bags of which had made their way across town from Angell Street to Beachmont.

Why cop an attitude against a miracle-like synthetic fertilizer? Who cared if the stockpiled poisons of World War II, created to defeat the enemy, were sold to farmers as pesticides and soil enhancers that would increase crop yields to feed the burgeoning human population? The chemicals were promoted as good for our crops because they were bad for the bugs that ate those crops. Unfortunately, the chemicals killed all the good bacteria in the soils, too. And so more soil enhancers and pesticides were needed. And more. Until the suspicion arose that maybe those synthetic fertilizers were doing more harm than good because they were depleting the soils, not rejuvenating them, such that in sixty years, soils around the world will be dead and unable to produce food.

That will be a problem, dead soil.

Also, yields of conventionally grown vegetables and fruits are leveling out. And studies of the effect of pesticides on humans verify this fact: pesticides were designed to kill.

It seemed to me those miraculous synthetic fertilizers no longer fulfilled the promise of the advertisements that sold them (no bugs, higher yields). The only promises being fulfilled were profits for the too-big-to-fail companies that sold them and a chemical dependence of conventional farmers on those companies for their patented seeds and fertilizers to the extent that tens of thousands of farmers in India, upon recognizing the vicious cycle they were in and seeing no financial escape, were committing suicide by drinking those chemicals.

Gardening used to be such an innocuous sport. I would watch my father weeding his flower beds, covered with sweat and dirt and happily humming to himself. It was him against the weeds, and he used his relative brute force to keep them at bay, a never-ending battle in a millennia-long war of humans controlling nature for their benefit and enjoyment. Had the ground rules of that war changed? Certainly the stakes had risen, along with the sea level and profits.

Whatever. I had seeds to plant. Our chickens and composting hay were all up at Darwin's View. I had three days to jumpstart the Beachmont garden that we had never prepped for winter, and now it was almost spring and, apparently, the Road Chicks idea had been lost in all our *planning*. Why, I wondered, were our hens not with me, earning their keep by scratching up the remnants of the lawn and prepping our garden beds? Instead, they were eating all the grass seed we had spread on the hill at Darwin's View, where we had no gardens because we might have a house up there but the site work all around the house had still to be done. It didn't look like it would be done for weeks, if not months.

Damn the torpedoes! If the hens weren't in Providence, and if Miracle-Gro was all that I had to plant in, so be it. I lined up the seed packets, all fifty of them. All organic. I imagined those cherubic seeds cringing, shrieking No! Don't Do It!

I did it. I took the organic seeds and stuck them into the light, fluffy depths of the Miracle-Gro seed starter mix, which didn't smell remotely like manure. It smelled. . . . I inhaled—a sense memory of my mother's art studio. Of the fumes that had lingered after a firing of metals shaped into sculptures. A quick outward snuff, and I shut down any thoughts that those two bags of soil were falsely labeled and filled with carcinogens. The seeds sizzled. I ignored my doubt as so many good American farmers must. We trust that the public welfare still overrides profits.

I wasn't so confident about my seeds' ability to overcome my gardening tactics. I attempted to send out positive vibes. If I, the overseer of those seeds/seedlings/plants didn't have faith in them, what would the seeds/seedlings/plants feel as they struggled into existence?

Crowded. Very crowded. If I put only one of those tiny-as-a-grain-of-sand seeds into that euphemistic soil, likely as not, it wouldn't grow and I'd have a bunch of used seed starter and no greens for Bessie and the girls. And does one compost Miracle-Gro? I put two . . . three . . . sometimes four seeds into each cell of the seeding trays. It felt more hopeful. More seeds, more chance one would sprout. The odds were in my favor that something might grow.

At which point I remembered that I was supposed to have written down what was where. Thus the bag of handy-dandy Popsicle sticks. I had bought them specifically for the purpose of labeling the seed pods, such that when it came to planting them, I would know which were pumpkins and which eggplants. And why was it that the harder I tried to be organized, the less organized I got?

Carl usually helped. He would come out of his practice room, looking a bit dazed as he reentered the world from his music la-la land, and study me. A hugely irritating habit of his, to stand watching me, exuding a heartfelt wish to help me out. He fully appreciates that I am trying my best. He knows that I have the will and determination. He also knows that the more flushed my face becomes, the more harried my hair, my brows pointing up in a devil's smile, the deeper into my struggle I go, the more reactive I will be to his suggestion. He wants me to be successful, and I take his words and presence as criticism because I know I'm doing it all wrong. For example, seeding. He gently, as kindly as he can, recommends a more . . . organized process. He might note that maybe I should slow down, be neater, pay attention—a hugely insulting comment that leaves me offended and tossing everything away because he is right, and why do I bother to do anything if he's going to come and do it all over again, correctly?

But in the basement of the Beachmont house, I knew that this time would be different. I hoped it would be. Carl had gigs to prepare for, and a bathroom to demolish. He didn't have time to hover and I could do this on my own. There was no good reason for me to fail, and it would be my project. I would have the pride of accomplishment, to be able to say I grew those watermelons and tomatoes from seed!

Having shoved the myriad seeds into pods and "seed starter," I left Mother Nature to do her thing. If the seeds grew, they would reveal their identities. As for me, I marched myself outside. I had no plan and promptly became mired in indecision. There were too many options. Which seeds should I plant first? Where? My frustration level rose. Are they buried in the soil or strewn randomly? Raked? Watered?

Over the next couple of days, I broadcast lettuce, arugula, beet, cucumber, kale, pea, and bean seeds. I pecked and scratched at the six raised beds and knew I lived in a fantasy. I wanted to learn how to be self-sustaining. I fantasized about growing an abundance of vegetables in flourishing gardens. So much potential and opportunity . . . and I ended up gardening how I seemingly did everything: too rushed, too dreamily, and then life's hurly-burly.

I dropped my trowel. If I didn't hurry up, Carl and I would miss the train to New York City, and it was no small wonder no one knew where Carl and I were on any given day. We had been based at Darwin's View for nearly five months, but were not on the radar screen of our New Hampshire friends because we had been in nesting mode since our arrival, not entertainment mode. Our Providence friends, meanwhile, assumed we were at Darwin's View, and were always hugely surprised when we were sighted in Rhode Island. And both groups of friends were correct in our/ my not being here or there when I was in New York City visiting my mother and/or Carl was on the road. Had nothing changed? We were hyperactively *doing,* still not *being.* And yet we expected gardens to grow and animals to prosper.

The Chicken Palace

LIFE AND DEATH, THOUGH NOT IN THAT ORDER

Back at Darwin's View, Carl and I gathered ourselves to steward the property. We divided the land around the house into three tiers. Tier One, the area closest to the house and to the south, would be some sort of Eden-like mix of (medicinal?) perennials and annuals and the site for the Chicken Palace, which was in the process of being built. Tier Two—a step down to the south from Tier One, and four times its size, would be where Carl would eventually plant twenty-five raspberry canes, and we would plant—with the requisite pine chips and compost—twenty blueberry twigs. And an asparagus bed, and a more fulsome, organic vegetable garden, all of which would happen whenever the site work got done. Meantime, the dirt was blown hither and yon, turning into muddy gullies when it rained because there was nothing resembling a plant to cling to. And Tier Three, the rest of the hilltop to the south and west, which had been seeded with a cover crop mixture, was ever so tentatively beginning to grow into a rough-and-tumble pasture.

Looking northeast from Tier 2

Across the driveway to the east was the newly moved, refurbished, and soon-to-be decommissioned Hay Chalet coop, and beyond that, acres of field. A barren, empty field and the house bumped up in the midst of it. My dip-of-the-toes reading about crop rotation and companion planting served only to add levels to the decisions about where to put what, and it didn't matter anyway, because nothing could be done until the site work got finished, and there was no sign of the excavator returning. Darwin's View exhibited our powerlessness in the world and my inability to effect change. Just thinking about the great outdoors and all things Darwin's View exhausted and overwhelmed me.

Carl, on the other hand, got energized. After a couple of hours of practicing, he would put on his work clothes and gallivant out to the hinterlands of the eastern field. There the suggestion of an orchard grew. We had planted the mostly apple trees—though there was also a plum tree and a cherry tree—five years before. All the trees, without exception, had shrunk, not grown.

Given past experience, our presence might not necessarily have instilled confidence in those trees, yet the cherry tree went so far as to provision us with a solitary, bright red cherry after Carl put fencing around all those treelets to protect them from deer. Had I listened, I might have heard a collective sigh of relief rise up from the orchard as Carl shoveled buckets of compost around the trunks of the trees and, glory be!, watered them.

96

Yes, Carl coddled those trees. My opinion on the subject was more flick-that-switch. This was survival-of-the-fittest land. If saplings got knocked down, whack, to the ground, I supposed staking them for support made sense, but water? I left it to the plant whether it would survive or perish. (I later discovered that I had quite naturally adopted permaculture farmer Mark Shepard's STUN technique: sheer, total, utter neglect). Thus, and in short, Carl took on watering duties, leaving me to collect the happy eggs.

Climate change skeptics had nothing over my own belief that our slightly feral chickens existed far away from such crass things as hunger, drought, and mortality. That's why I had bought into chicken steward-ship: to keep me wading in the fluff and cheeping of denial. The chickens, however, had apparently decided to drag me onto that flagship called reality. On this particular day, May 13, 2013, we had four big, beautiful eggs, which meant all the hens but wee Lola had laid one.

"Beautiful!" I exclaimed, picking up one of the eggs from the nesting box in the Hay Chalet. "Thank you, girls! You certainly know how to lay an egg. Look! It's so brown and oval! It's perfect!"

They ran over to admire their work—except Lola, who had chased a grasshopper out into the field and now scratched out there all by herself. She paused in her busyness to look toward us. Perhaps she wished she'd had the forethought to lay an egg, too, so as to be part of the grati-tude-and-love fest. She appeared to sigh, then returned to her work. She was such an intrepid little hen, braving the vast distance between herself

and the rest of us, the possible hawks soaring in the sky and fishers lurking in the grasses—risks I, myself, hesitated to take because I was not a brave chicken. I determined that she de-served a treat, too.

"Yellow cheese!" I called as I headed to the house to put down the eggs and pick up the plastic container I'd filled with treats.

Lola

Lola came running, her legs aided by a grand flapping of her wings. Chickens are really cute that way: they can't walk to a treat fast enough and so half-fly to it, the head thrust forward to get there first, the eyes bright and excited, ready to inhale whatever is offered. There is no shame in their frenzy and greed. Unless it's not quite the thing, in which case what embarrassment exists falls on the person who'd raised their hopes only to crash them down.

This was not the case with yellow cheese. Yellow cheese is in the realm of mealy worms and, therefore, worthy of hullabaloo. Thus, I was surrounded by my girls and Big Red when I held out the little squares of yellow cheddar cheese that I had cut up, just for them. All those cheerful, beady little eyes adoring the cheese, if not me. How could I not smile? Any anxiety that the heat of the day might be unnatural, the lack of rainfall, too, disappeared in those moments of cheese offering. First to Big Red, of course. One mustn't insult the resident roo! He daintily took the piece of cheese from my hand, dropped it to the ground and excitedly clucked his "here's a treat!" guttural sound. Repeat. While he laid claim to the credit for the cheese, I counted the happy, spoiled hens. And then again. Odd. I came up one hen short. I went through the names and noted the absence of Chipper. "Big Red?" He avoided meeting my eye, acting exactly as he had the time a hawk had hovered and he had reached the coop before any of the hens. "Where's Chipper?"

Instead of admitting he hadn't been paying attention—something I could hardly fault him for, given my own *modus operandi*—he grabbed another piece of cheese.

"Here's a treat!" Four hens, not five, but isn't he an important fellow? I saved the last of the cheese for the missing feral girl. Where was she? I began to look for our black and white, who resembled a stained glass window with a rose comb, hen. She wasn't on Tier One or Tier Two. I approached Carl, who had returned to a more valid task than watering trees, namely digging a ditch, like a moat, around what would be the Chicken Palace and its runs.

"Have you seen Chipper?" I asked. He paused to wipe the sweat from his brow. His face streaked with dirt. His dimples dimpled because he was in his element. Dirt. Sweat. Activity.

"Nope," he replied.

Chipper

I nodded, looked around the yard. On the porch. Back over to the Hay Chalet. Was she in the run? Indeed, she was—on her side, in full *rigor mortis*.

Remarkably, I was fine. Perhaps surprised. I had known her since she was but a chick impersonating a chipmunk. Thus, her name. She had been a happy hen. The one I occasionally called Jonatha because she looked and sounded like a seagull. Like Jonathan Livingston, she was the quiet, solitary one. I always thought of her as a thoughtful chicken, looking at the world and wondering about it. The why and the how. She was the middle-of-the-pecking-order one. Our Wyandotte.

I picked her up, a stiff chicken with such soft feathers. I'd never appreciated her beauty, having been distracted by her mind, not her body. Sweet, matter-of-fact Chipper.

"Carl," I said. My tone of voice must have caught his attention. He looked up, and his posture sank.

"Oh, no." He set down his shovel and came over to hug me and the chicken.

But I was fine. Because so what? One bird dead. Who cared? I called our vet. One might have died, after all, but what if what killed her were contagious? What if all the chickens died? What if we did? Our vet agreed to do an autopsy, which is, remarkably, what happens every day in homes around the world: the cutting open of a chicken's breast.

Not in my home.

I didn't watch but imagined the knife diving into her black-and-white feathered breast, the flesh rended, her feathers' stained glass reddened with blood, soft, mortal, not shattering into shards. And I was fine.

A few hours later, the vet called with the results of the autopsy. Then I cared. Sitting in my office, weeping, I felt betrayed and offended. I curled into myself, not wanting to be near anyone, talk to anyone, least of all the chickens because it was the whole mortality thing. Life's meaning. One's place in the world. Why *are* we here? Why are some so lucky and others so not? Why did a ten-month-old hen die of a liver that was mottled into what looked like rotten cottage cheese, according to the vet? The cause of death was cancer and, thus, not contagious as I had worried, unlike the remaining chickens, who pecked the ground as if nothing had happened. Just so us. Just so me. In the face of mortality, yet we live.

We buried Chipper as the sun set. Carl dug a hole next to the big rock that faced to the west and Mount Monadnock. He placed her in the hole with as much grace and respect as one can put a chicken that's been sewn together after the ravagement of an autopsy. While Carl played "Taps" on his trombone, I shoveled the dirt into the hole. We added stones to mark the spot and to be sure no nocturnal creatures could unbury her. Panda and Big Red oversaw our actions. When we were done, they led Ping, Lola, and CooLots in a parade down the hill and began to inspect Chipper's grave for any interesting worms or bugs.

"Sad," I said, my chin trembling. Carl put an arm around my shoulder.

"She had a good life," he said.

"I guess so."

Our first death. It was odd to have been counting to six for ten months. Now only to five. Again and again. I didn't want to miscount. Lose another. Who would die next?

Twenty years prior to Chipper's death, almost to the day, my mother, my sister, and I stood at my father's bedside in our home in Stonington, Connecticut. He had moved into the guest room off the kitchen a couple of months before, with the intention of recovering from his bout of cancer. Unfortunately—as the doctors wrote him a few weeks into his "recovery," and in response to my father's inquiry whether another operation might be in order, given he was losing feeling in his feet—there would be no such thing. The nine operations he had endured notwithstanding, another operation would be pointless. The cancer had metastasized and would kill him. It fell to me to read my father this fax.

Fortunately, he had that bright, sunny room with its big greenhouse window. He could look out from his hospital bed at the bird feeder.

Chickadees flitting in, looking about, dipping in for some seed, and off again as the brilliant vermillion of a red cardinal approached and tufted titmice performed their acrobatics. And if, in his distraction and thoughts, his gaze followed the chickadees, red cardinals, titmice as they flew off, free and healthy, across the expanse of the hardy, slowly greening lawn, he might have considered his years of mowing, of pulling up weeds from the dirt, and of planting his beloved dahlias, gladiolas, bearded iris. And the days and weeks passed. He asked for a copy of Proust's *Remembrances of Things Past,* in English as his concentration wasn't what it used to be. And now it was Tuesday, May 8, 1993, and Carl was there, too, if in the background. He was preparing to go on the road. The band had already driven off in the band bus to start a two-week road trip, driving the three thousand miles to their first gig in California. Usually Carl would be with them. He was the driver *extraordinaire* and bus fixer. But this time, I needed Carl. And so he was on the phone with a travel agent, buying his plane ticket, while I held one of Dad's hands and my mother held the other. Or was Xanda holding his hand and I standing at his feet? In any case, the three of us were with him. His trinity.

Someone once said that Dad's list of priorities was first, work; then, Mom; then Xanda and me; and way at the bottom, the rest of life's stuff.

We already knew he was dying. Dad had dictated his obituary the previous week and I had typed it up. He had told us what to do with his books. His will seemed in order. He wasn't overly "uncomfortable," our family's euphemism for pain.

His sisters and mother? "Get thee to thy gardens," or some wonderfully dramatic directive like that, is how my father said good-bye to them. He was quite pleased with the effect, especially as he lived for another few days. It provided a sense of placement and control to think of his siblings and mother in their gardens, weeding and pruning and preparing for the summer's abundance. We were all in our places and now, after all that time of his *dying,* the hospice nurse told us that he was *actively* dying. Certainly, he was unconscious. Upon hearing that his death was imminent, his trinity—Mom, Xanda, and I—agreed we should call the church to reserve a space for the weekend. I put in the call. The minister's voice toned down to a melancholy level when he heard my request.

"Ah," he said. "My condolences. When did your father die?"

101

"Oh, he isn't dead yet," I replied. A long pause. I clarified. "It shouldn't be long now." A yet longer pause and so, to avoid any misunderstanding, I said, "The hospice nurse says he's actively dying."

"Why don't you call me back," the minister said, his voice toughened. "After the fact."

I returned to the bedside. Carl left to go buy his plane ticket. The clock ticked. My mother—still active and healthy—removed Dad's hearing aids. "He won't need these anymore. They caused him such pain."

Emotional, physical, psychological. So many kinds of pain, and I am sure that, at this point, I was holding his hand—his left hand, because I remember clearly that I spent his last minutes watching the artery in his neck pulse. I thought, *When it stops, he will be gone.* He breathed in, then out of his mouth. It hung open. His patrician nose. Bald head. The pulsing slowed. I thought, *I am not ready for this.* I told myself, *Things will never be the same.* I asked, *How can I stop this from happening? What will be left?* And is the past mere nostalgia as the agony of keening dwindles to heart spasms that only occasionally flare into hurt, but not enough hurt to cry? He had been sick for years and on the downward slope to death for months. Months of tears, and waiting, and hoping, and watching someone die and nothing to be done. The heart goes back to the past with longing, and the future? There is no more future. And what's it like to die? The relief from pain opposing the panic of wanting another chance.

Dad had called me into his room the previous week. I was on night duty. It was around two, dark and chilly. I wrapped myself in a blanket and sat in a chair next to his bed. He rambled on about a water shortage, how it would happen in the Northeast, not the Southwest, and I had to get in touch with my Uncle Ash. I wondered what Ash had to do with water shortages, and chalked the absence of my father's usual intellectual lucidity to morphine-induced hallucinations. He wandered on. There would be a war, and they would stop the flow. There would be no water south of the Carolinas, and on and on until he stopped to ask if I had gotten it all down.

"What?" I replied, sitting up from my doze.

"Did you get it down?"

"What down?"

"The outline! For my book!"

Uh, oh, I thought.

"No?" I said, reaching for a notepad and pen. Apparently, he had been narrating the plot for a mystery novel. Dad had loved to read Dick Francis, Agatha Christie, George Simenon, and other such mystery novelists, like light snacks between the heavier meals of history and political science. He now had time to write such a book. Why not? And so he had composed a mystery novel as he lay on his deathbed. Prescient, he knew water would become a rare commodity, one people would fight and die for. Thus, my Uncle Ash. He was in the publishing business.

When I admitted I had not written anything down, Dad was hugely irritated with me. What could I say? I felt as if I had failed, and not for the first time, his expectations. Plus, I was familiar with his frustration. How many times had I dreamed a book, fully outlined, needing only to be written, but as I got closer to consciousness, the ideas raced away like chickens eluding capture?

My father and I sat in silence. Frustrating, yes. Most of all because we both knew he wouldn't have been able to write the book anyway. He didn't have the time, and yet he had hoped, as the clock ticked and he became weaker. Dipping into unconsciousness, then waking up. Dipping down, we cling even though we have no choice but to let go. We hold tight, hoping we might stop . . . but it has already happened. The artery in his neck stopped pulsing and we three, my mother, my sister, and I, were left. Our hearts beat on.

You think you won't survive. You aren't quite sure how you will cope. And then you do, even if you don't want to. And then?

I called the priest back. Carl flew to California to make his Lake Tahoe gigs; he would return on a Saturday night red-eye flight, arriving in time for the Sunday afternoon funeral. Life proceeds. If we keep breathing, going step by step, not looking too far ahead.

My father then as patriarch—mind over body, the sun, expectation, self-worth and acceptance and hope. My mother now as matriarch—creativity, the moon, the womb, nurturance and strength, healing. My parents, as with the bigger world and its species, are gone or irreparably changed. Without them to fall back on, I reached out. What could I cling to, hold, as I pulled myself up the mountain?

Four hens to one roost-
er is not a good balance.
Certainly Big Red was
of that belief. The more
hens, the merrier for
him. The consequence?
When a cousin of Carl's
needed to rehome two
of her Barred Rock pul-
lets, we adopted them to
replace the irreplaceable
Chipper and to alleviate
the stress of Big Red's re-

Daffodil and Chickadee

lentless attentions on Ping, Lola, Panda, and CooLots. But Chickadee and
Daffodil did not suffice. Worse, the pecking order was a mess. The two
City Girls were big and aggressive. Ping was no longer on the top rung,
chased as she was by those two larger hens. And little Lola was harassed by
all the giant birds around her. I suspected she was developing an inferiority
complex, an ugly sight in a bantam hen who used to be such a forthright,
spirited chicken.

This, along with the horrors of CAFOS—concentrated animal feeding
operations, known, too, as factory farms—and desertification and over-
population, gave me something to think about when awake at night and
early in the morning: the next generation.

I was not alone in this line of thinking. Panda, in all likelihood to avoid
the embarrassment of Big Red's flirtations with the new hens, joined me
by settling into her nest.

"Broody" is something I have never been.

. . . Actually, I do tend to brood. I sit in my office and frown and
peck irritably at the keyboard, contemplating life and death and purpose.
Which is, I suppose, what a broody hen does while she sits on the eggs
that she will hatch. With "will" meaning her will against the world's. Her
existence—and possibly those around her if they get in her way—is at
stake. That must be why it is so hard for me to say no to a broody hen,
specifically Panda, who, from being the gentle queen roosting in her place
next to Big Red, became a rather moody, unfriendly creature. Even Big

Red learned to leave the hissing, fluffed-in-anger, stalking-about hen to her eggs.

Four of them. I let her keep four eggs.

For our part, Carl and I headed to Connecticut for my annual family reunion week. Ezra Landis, one of our neighbors' sons, was ready to move in while we were gone. The chicken thing was out of my hands in pretty much every way, except, perhaps, the responsibility and consequences ways.

We left. We came back. Twenty-one days to the day, I heard them. Cheep, cheep, cheep, cheep. Our Mutt Chicks huddled under Panda, whose eyes were weepy with joy. Pong, Claytonia, Beatrice, and Cordelia, the last bearing a remarkable resemblance to Big Red back when he was Rhoda: relatively enormous, red with beautiful eyeliner markings. If she turned out to be a he? Cornelius.

Did the thought of a baby roo distress me? Yes. But I figured the chickens would figure things out for themselves. "Chicken rules," I reminded myself. And "Don't let them lead you by the nose," Carl reminded me as I stepped outside for a visit with them.

Whatever my mood might be when I walk out to the chicken coop, upon arrival, my attitude shifts. My heart lifts with a contentment that I typically feel only when hanging out with our cats Nick and Nora . . . and pretty much all other animals I come into contact with, except for humans and snakes. And wasps. Ticks.

The point being, when I have the opportunity to be with a soft, potentially cuddly animal, I am. With the chickens, I existed, and watched them as if, by osmosis, I might take in their acceptance of life, their beingness when sprawled in devotion beneath a ray of sun, or when in corpse pose while taking a dust bath. Excepting when the pecking order revealed itself, the chickens brought me great happiness. I felt useful when I cleaned the coop or when Ping ran to me, full of the anticipation of the treats I might have for her, or when Big Red chortled excitedly, offering up to his hens the fat grub he had found. Whether I fully understood the depths of the chickens' teaching moments, I knew that I needed more of them. Thus it was that I had asked—weeks and weeks *before* there were any signs of the City Girls and Mutt Chicks—a local Ameraucana breeder if we might buy two of her pullets. We adopted the girls, Neurotic and Terrified, shortly after the mutts hatched.

Neurotic and Terrified

Carl didn't know I had Darwinian plans. Neither did I. I had no idea the amount of kerfuffle that would occur in my heart now that we had thirteen heartbeats to care for.

Counterintuitive, to claim, on one hand, that they helped me to *be* while on the other noting the heightened anxiety level? But *life* caused me anxiety. Why would chickens be exempt from the *sturm und drang*? Example: chicks are totally adorable. Their cheeping, however, is one of the most grating, anxiety-provoking sounds I know, proving that joy counters angst just as love counters hate. The negatives might seem to dominate, but I wouldn't give in or let go of joy and love. With chickens, I could take the necessary risks.

Thirteen is known to be a bad-luck number, proven by the fact that our chickens did not have a 50–50 birth ratio, boys to girls. We had a 75 percent return of boys. And three separate flocks: the Mutt Chicks and Panda in the nursery; Neurotic and Terrified huddled in the dog cage in the run; Big Red and his hens in the main coop. Although the Hay Chalet had been upgraded to the Chicken Palace, it was too small to house them all. We would have to create yet another chicken coop. Before winter. It was July. We had plenty of time.

"Whose bright idea was this, of chicken hatching and adoption?" I asked.

No answers. Only tap, tap, tap. Typewriter keys? A cane knocking on the door? A long, bony finger on the shoulder, a reminder that life and death walk hand in hand, too.

Lola

NATURE AT THE DOOR

Our valiant Lola disappeared, leaving a pile of feathers. The next day, Neurotic evaporated. She left nothing behind except her sister, whom we renamed Lo, for the Lonely One, at the bottom of the pecking order. I noted that it was raptor migration season and locked the chickens in their run. This did not make for happy chickens. They paced and clucked and pressed their faces against the fencing. Bugs! Worms! Acres to explore! They wanted out—but only because they had no idea of the danger. If they knew, I was sure they'd be quite content to remain in their safe area. I shut the curtains of my office so as not to stare out, and repressed the stress of their complaints by reading more about global warming, which was actually climate change, which was really climate disruption and weirding. Not long after that, Carl announced I had lost my sense of humor.

A devastating comment. I had only just found it. After years of my writing being on the too-serious side, I had learned to add a dash of fun. I could find humor in amateur gardening and chickens, snowstorms and our on-the-plump-side cats, but how to find humor in the death of our planet and a past that became more distant and irretrievable every day? And yet, my spirits were oppressed not by the fact that I would never experience a days-long migration of passenger pigeons darkening the sky, or that the water and air and soil of the earth had been ravaged by human extraction for money rather than survival. Those were things done and gone that made me terribly sad but that I had no control over. The past that I yearned for was my entirely accessible home at Beachmont Avenue in Rhode Island. How was it that Darwin's View had become our default roost? At this thought, Big Red would invariably crow in the background.

So? Why did I not just drive to Providence for a dinner with my girlfriends, easy peasy? Because I'm not good with transitions. The idea of driving to Providence sparked a deeply unpleasant anxiety that had no apparent bottom. Car crashes. Driving off a bridge. Nothing was out of the realm of possibility. And what could happen to the chickens and cats while I was gone? A fisher in the coop. Nick might open the porch door— he is Houdini incarnate—and proudly prance outside, only to be tackled by a hawk. Fur and feathers everywhere. And Nora, seeing the open door, might go to meditate in the forest, looking, as she often does, like a furry, marmalade Buddha. She with her soft, tufted paws creeping about, seeking out mice to play with and eat. Her golden eyes wide and soon enough afraid, because she would be lost and without the beta blockers we gave her to keep her faulty heart pumping. What if Carl hurt himself, as he is wont to do, and I wasn't there to drive him to the hospital? Then what?

No. Stressful enough to go on a day's errands in Keene, the nearby small city, but all the way to Rhode Island, and there to remember what was no longer? Which maybe was the deeper reason for my staying in New Hampshire: to avoid the rending of my heart when I would go back.

A puzzle. I juggled the two places in my heart. Obsessively ruminating on what I couldn't have, and doing nothing to get what I could, exposed the fact that my childhood patterns yet ruled my behavior. Just as at boarding school, I had learned to pay no attention to the heart and soul; thus did I scoff at my heart's desire to return to Rhode Island, numbing it in favor of the idea that I might do something at Darwin's View. In my own mind, if nowhere else, Darwin's View began to represent what I knew had to happen in the world writ large: the discomfort zone of transition from the past to the future.

108

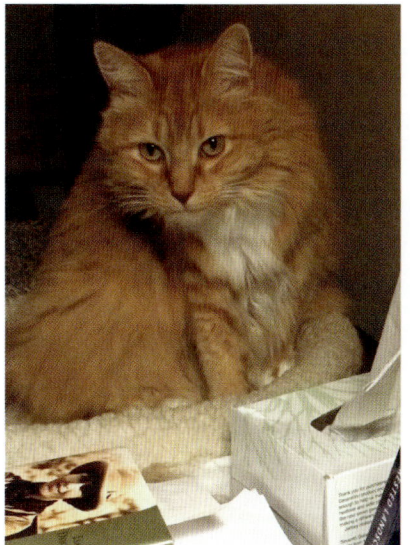 

Nora   Bobcat

I did not trust nature, and resented her relentless intrusions: Wind. Flock decimation. Time's passage and no control. Further justification for my frank outrage at life in the form of Mother Nature arrived one morning. In the midst of a heartwarming fantasy about Lola returning, her golden chest puffed with pride by a line of tiny, peeping chicks behind her, I noted a rather large Siamese cat at the coop door. The morning's mist had yet to rise. Through that damp veil, I compared the cat outside to Nora, whose nose was under a tufted foot as she snored obliviously on the porch. Her furry cheeks bore a remarkable resemblance to that other very large cat's tawny cheeks. And I knew that Nora's bleary-eyed demeanor when disturbed from a nap could take on a slyer, more predator-like mien when stalking mice and moths. Yes. Nora and her fellow *Felidae* proved how far two related creatures could get from their roots. The sky brightened. The four-legged creature pawed experimentally at the run door and morphed from soft, cuddly Siamese-looking kitty to—with the growing cacophony in the coop—bobcat-devil out to eat the chickens.

I yipped and hollered for Carl, who was still in bed. I ran out to the screened-in porch, past the dozing Nora. Pulled on my rubber boots. Grabbed the handle of the sliding door. Paused to consider bobcat psychology.

Cornelius, Clayton, Pong, and Beatrice in their adolescence

Are bobcats the small, scurry-away type of predator? Or do they go into attack mode when approached by a shrieking, two-legged creature, waving its arms and perhaps a broom? Would the bobcat chase that two-legged creature back onto a screened-in porch, perhaps through the screens, pausing for a tasty morsel of fat house cat?

Not in the mood to finally have an answer to one of my questions, I returned into the confines of the house, carrying a disgruntled Nora. There I found Carl in his birthday suit with binoculars and camera on the ready as the big cat strolled leisurely down the stone wall, looking back at the coop every so often as if purring, "I'll be back."

Nature at the door is uncomfortable. Porcupines threatened the tops of apple trees and voles the bottoms. And then there are groundhogs, a.k.a. woodchucks, who begin as such adorable, wee schnookums but, within weeks, grow into evil little bastards eating one's lettuces and pretty blue flowers. How dare they?

I guessed it had something to do with hunger and survival. Humans had taken over so much of the world that the wilds sometimes have no option but to trespass. Human boundaries are not nature's. We had passed over hers without a passport or respect and, as I knew from watching Parkay commercials as a child, it's not nice to fool Mother Nature. Nature is tough. She has rules. One slight misstep and it is natural selection time. A brutish existence, as Lola and Neurotic had proven. Nature's rules crushed denial. So did our three cockerels, who grew bigger and more gloriously male with every passing day.

Pong, Cornelius, and Clayton, just by their existence, were troublemakers, strutting about, followed by their doting little sister, Beatrice, she being the exception to the rule that no one wants mutt cockerels, however dapper they might be. The day was fast approaching when Big Red would kick out his three sons. I pictured the boys, their belongings wrapped in bandanas and tossed over their shoulders, dragging their spurred claws down the driveway. Their bright expectation, confidence, and friskiness subdued. Panda and I would wave good-bye, dewy-eyed because their future, by most standards, was grim.

People don't have much respect for cockerels. They nonchalantly point to a soup pot, and say, "There are worse things that might happen to them: Bait for cockfighting. Live plucking. The chopper." Others suggest "forgetting" to put them in the coop at night. Evisceration by a fisher or raccoon. Problem solved.

They *say*. But if a fisher or raccoon showed its furry face? Not only would the current problem not be solved but a new one would be created. Leaving our cockerels out of the coop would *improve* their innate survival skills. (Sleeping in trees. Manicuring their spurs.) But if, by chance, the boys didn't send a predator scurrying for its life? Then that fisher/raccoon would know how good chicken tastes and where to find more. Thus, leaving the boys out to fend for themselves was not just heartless but counterproductive. *It was baiting the Chicken Palace.* This being food for thought during the happy happenstance that the bobcat's visit had caused: Carl was fully on board with the Fort Knox Coop concept, and motivated. We picked up nature's gauntlet and contemplated the next rendition of the chickens' home: Chicken Paradise. We would keep nature at bay, come hell or high water.

Chicken Paradise in winter

A COLD NIGHT IN HELL

Chicken Paradise was the sixth rendition of a chicken coop that we had created since we adopted our six chicks. In winter, the hay in the runs was a foot deep, and at night, as it was on this February night, there was a warmth and comfort in its confines. It wasn't unusual for me to pause as I closed and secured the doors. The rustling of the hens settling down for the night was soothing. In the darkness, with the full moon outside and the temperature hovering around nine degrees, I took in the sense of peace and innocence that would soon end. I counted: seven hens, four cocks, three death warrants.

The power of death was not what I had in mind when I adopted our original chicks. Nor had brandishing a pair of tree-branch loppers over a chicken's neck entered my thought process when we let Panda go broody and hatch that clutch of four eggs. Even when Big Red's daily solos had made way for a Four Cock Choir, I had hoped. Instead of worrying about the frigid . . . no, balmy . . . no, frigid weather, I had reveled in my role of chicken lady and called farms and animal sanctuaries, seeking happy homes for our boys, who perked up the day and kept things at a basic level: food, sex, sleep. I had even approached our vet about castrating them. No luck. They were too old by the time I asked, and my sense was that our vet, who had been willing to dissect Chipper when she keeled over dead, drew the line at castration. Though it's done all the time. That's where capons come from. Castrated cockerels. Usually without anesthesia. Uncomfortable thought. But with or without that numbing effect, the answer was no.

I had no intention of being party to the murder of our Mutt Chicks, and decided to keep them. I had learned from a farm animal sanctuary that cocks can live together harmoniously if they have enough room and hens. Ten hens each.

We had the room. One hundred and ninety-three acres, twenty of them on a sunny hilltop. A chicken's dream, if one subtracts the hawks and bobcats. The only problem was the heaps of snow that kept falling, thereby narrowing the acreage down to about six hundred square feet.

Even I drew the line at adopting another thirty hens.

Days, the tension in the coop was high: Big Red studied his sons challenging each other, not him. Testing their vocal cords, crowing in dissonant harmonies. Cornelius, in particular. He was a young cock and getting bigger than Big Red, who, at one and a half years old and with a frozen comb, looked positively ancient. The hens huddled near him because he

Big Red

The boys in their glory (left to right) Cornelius, Pong, and Clayton

113

barred the boys from easy access to them. But he couldn't be everywhere, and too often the boys would do what cocks do to hens. Feathers everywhere. The hens appeared frazzled and exhausted by all the amorousness.

And then came the morning I entered Chicken Paradise to see Cornelius, proud and upright, his throat glistening with blood. The hens clucking and nervous. Big Red not even bothering to dance. Muted. The boys' energy

A bloodied Cornelius

whirled as they chased the hens . . . but not each other. The next morning, Pong bore the signs of a fight. The coop and hens were all flecked with blood that had frozen a deep pink. Whether it was the boys fighting or Big Red maintaining his dominion, we had a problem. The flock was stressed. So was I. My three cockerels were only doing what they do and, thereby, were forcing my hand, if unintentionally.

Neither Carl nor I wanted to kill our boys, but both of us felt it had to be done, and that we had to be part of the process, if for slightly different reasons. For his part, Carl felt he should assist because he does eat meat. We had raised them. It would be appropriate to see them full circle, to the dinner plate. It felt natural and right.

For my part, I wanted to support my Mutt Chicks. By being present at their deaths, I might assist in their passage. I didn't know quite what that meant, but I did know that it felt important to be there. To recognize them as living beings, not things. I wanted to hold true to something bigger than my comfort level, because, once again, here was life. Here were the consequences of my choices.

Not feeling entirely up for the task that was before us, Carl and I walked down to see our neighbors, Hugh and Terry Landis.

In superficial ways, we hardly knew Terry and Hugh. They were real, live off-grid farmers. I imagined their lives being utilitarian, uncluttered, simple in the way that people with their priorities straight live. With our every interaction, I learned from their gentle touch on the earth, their gratitude for and belief in its spiritual beauty, and their ability to walk barefoot on gravel, and through fields, without fear of snakes or prickers. I held them in great esteem. Their presence in our lives somehow validated whatever it was I sought. They believed the world could be a better place . . . even if chickens did sometimes have to be killed.

By the time we reached their home, I was headed into full panic mode. The very idea of the two of us having to kill our three cockerels by ourselves was beyond imagining. We explained the situation to Hugh, with me feeling naive verging on stupid for having put myself in such an untenable situation. What had I been thinking, letting that broody hen get the best of me? Never, ever again.

Being the amazing neighbor that he is, Hugh offered to help us cull them.

Note: *to help us.* Not *do it for us.*

Intention. Responsibility for our choices. Sacrifice. Life and death. The boys were barely six months old. What right had we—had I—to take those lives?

My human right. Humans do this. Create and destroy. It was all so counter to what I had intended.

And so. It was night. The chickens were sleeping in the dark of the coop and I felt sick at heart as I faced this question: Who would die first? The heir apparent, Cornelius, our rose-combed beauty? Or Pong, the Barred Don? Or Clayton, the lowly roo, CooLots' son with his feathered legs, who ran from any hint of aggression, hardly daring to cock-a-doodle-doo, though the most aggressive in proving his point on top of a hen. All three protected their little sister, Beatrice. What would she do without them?

That not being the question.

While I stood in the coop, Hugh and Terry; Carl's niece, Sarah; and Carl stood in the driveway, a pot of water heating to 180 degrees. A pole, a rope, a T-shirt, and a pair of branch loppers. They waited for me to choose. It was very cold out. I would only be doing what happens every day all over the world, and in far worse form. I could, at least, give Cornelius, Pong, and Clayton a humane death, which was better than cock fighting. Or all those billions of chickens living in tortured conditions in farm factories, who then die miserable deaths. But, still, I asked myself as I shivered with something other than cold: what was I doing in a chicken coop on top of a snow-covered, windy hill, far from Providence, and living a life so beyond my comfort zone that I would initiate the killings of three sentient creatures that I wanted to let live? I wanted them to live and how dare life force me to do such a thing?

I debated which chicken to take first, and zeroed in on Big Red. He eyed me with no small amount of suspicion. Prescient? If I killed him, got the worst of the killing done first, we could move back to Providence. The option was there. I could choose to kill all the chickens, for that matter, be done with this little experiment. No more waking at 4:00 a.m. to Big Red's crowing. No more relentless worry of if they were watered, fed, healthy. Safe. Not frozen, or overheating. We could return to the familiarity of Beachmont. I'd settle into the house. Stretch in all its roominess and relish the familiar things. The wealth of stuff that, at times, was too much. Reverse acquisition appealed. And one morning—I knew this, even as I touched Big Red's back, gently, his blackened comb—I knew if we moved back to Rhode Island, one day I would stare out my office window at the park and miss him. All of them. The chickens had taken up a space in my heart. And, clearly, they still had lessons to teach me.

I reached . . . for Pong. We named him for Ping, not realizing he would not grow a rose comb à la Dominique but a single comb à la Barred Rock. He was a Barred Rock male because the bars are sex-linked. So he was

Pong

one of the City Girls' sons. And quiet. He was surprised by my touch but didn't resist. I held him close because as we stepped out of the coop into dark cold, I didn't want to slip and drop him. I wanted to protect him, keep him warm. Comfortable and unafraid.

As we approached the wood shed, lit up by the Subaru's headlights,

I sobbed. I couldn't stop weeping because I related too closely. Is that wrong? Was it mere anthropomorphism to give Pong my fears? My wish to live? Images of the French guillotine and no way to escape as I handed Pong, his feathers damp with tears, to Terry. She cradled him, rocked him slowly, lowering his

Pong and Clayton

head. Blood to his head, relaxing him. She spoke quietly, entirely in tune with my ambivalence, and in deference to the process we were initiating: the ending of a sentient creature's life.

With gentle steadiness, she handed the very relaxed but alert Pong over to Hugh. Pong, with bright eyes and curiosity, watched as Carl helped Hugh wrap him in a T-shirt, ripped apart to more easily encase him. A papoose. Things took on a more ominous mood when they took Pong's legs and tied them to a broomstick that bridged over two benches. It was other worldly, seeing Pong hung upside down, lit up by the lights, puzzled and looking around him but calm, unknowing as his head hung down into a plastic bucket. And then—how crude and unromantic—a pair of loppers.

I turned away. Rebelling against the event even as it was happening. It was wrong. I didn't want it to happen. I couldn't look, and don't remember if I ran before or after the loud squawk! as Pong's neck was cut. But I did run, ears plugged against the noise of Pong's body, now irrevocably disconnected from his head, convulsing against the side of the bucket, nerves reacting, adrenaline coursing, his life's blood bleeding out. I ran halfway down the driveway, weeping to no avail because it was done. There was no going back.

People kill chickens and let them race around. For some, this is comical, seeing a headless creature running about in its death throes. People claim that proves that chickens don't feel. They can still move about without a head. Stupid chickens.

During the French Revolution, when all those people were guillotined and the heads were held up post-decapitation, some claimed the beheadeds' lips, their eyes, still moved.

Every year, billions of chickens live and die in misery. Why cry for Pong? He'd had a good life and a humane death. Better than most humans can hope for.

And yet, I wept.

Maybe for lost innocence. Maybe for a kinder, fairer world. Maybe out of dread, because we still had two more to go.

When I returned to where the others were stamping their feet, clapping their hands, trying to get warm by the pot of hot water that would be used to dunk and loosen the feathers, I felt sickened and ashamed. I had called Sarah earlier that day to ask if she would want to join us. After all, she has a unique sensibility, collecting skeletons and insects, creating cabinets of *wünderkammer*. And she was always taking her daughter, our Peyton, off on eclectic road trips to Chinatown, Canada, Rindge. Also, she had been kept up to date on all things chicken at Darwin's View and had, herself, helped with chicken care, not least the hatching, growth, and revelation of our cockerels. I thought she might be interested in just such an event as we faced, thereby ignoring the fact that she had recently decided to cut back on eating meat. That she might be of the ilk who preferred not to *see*. That, though she found the bodies and skeletons fascinating, she had, in fact, in my vague memory, expressed queasiness around slaughter.

Thus, the pause after I asked her if she'd want to come up to see the killing of our three cocky youths.

"Do you need the moral support, Tory?" she asked, cutting to the chase.

"Yes!" I said. Unbeknownst to me, that was exactly what I needed, and she had come, and I had deserted her, thereby failing not just a chicken. Whereas the others had stood by, there for the transition from life to death, I had fled. I had hoped to be braver than I knew myself to be. I hadn't been.

Pong's body had stilled. We went inside the house to warm up with a cup of tea, pretend everything was normal, allow time for Pong to bleed out into the bucket. After not long enough, Hugh noted the time and that we still had work to do. While the others re-donned their attire, I headed back to the coop, already panicked and crying useless tears. Drinking tea had not been an option for me. I was too repulsed by the violence already done. To my core, I knew this was not right. I did not want to continue this very unequal hunt. It was a hunt of sorts. Only the prey was asleep. It wasn't fair. The chickens were helpless. I was helpless against the forces of life, and wondered at what point in my story had that night become inevitable.

And then I corrected myself. It was not inevitable. I was allowing this to happen. I knew it was my choice, even as I stepped into the coop.

Cornelius. Then Clayton. I ran away each time.

Later, in the warmth of the kitchen as we watched Hugh prepare Pong's body, I regretted that Carl and I had not been more organized. Even with the slaughter of our three cockerels, we had not *planned*. How much more natural and right it would have been to share our organic, homegrown, happy—well, they had been happy once—chickens with friends and family? Or, at least, to bury them in our chicken cemetery.

The wind chill outside hovered in the single digits, it was eight o'clock at night, and we had all been through the ordeal of lopping off three chickens' heads. That process takes something out of a person. Thus, we only kept one. Cornelius and Clayton went with Sarah, who had connections to a Chinese restaurant.

While still outside in the by-now obnoxious cold, Carl and Hugh had defeathered Pong's body. Dunking it into the hot water, they wiped away the feathers, plucked them off, occasionally with the help of pinchers. Pong—curious, gallant, and energetic—was gone, leaving a bald chicken's body, dimpled where the feathers had been. And scrawny. You would never find such a chicken in a modern-day grocery store. No oversized breasts; no meaty thighs. This was a lean, organic chicken body. Hugh sliced off its feet and placed them in the sink. I made a mental note to thoroughly vinegar and clean the entire kitchen. A natural teacher, Hugh sliced and pulled and cut the various body parts, identifying the intestines, the gall bladder, the heart—which was dwarfed by the testicles, thereby exposing the reason a cock's priority is mating and fighting. And then there was the liver. Hugh said we had to experience eating fresh chicken liver. He was disappointed that we had no butter. Olive oil would do. A fry pan. Flames. The liver sizzled in the oil and juices. Carl said it was good. I did not make an exception to my decades of "won't eat what I won't kill." Instead, I suggested to Carl that we go out for birdbath-sized martinis. At Bantam Grille's bar, with our bartender Morgan to listen to my woes, other customers agreed: they loved to eat meat and, sure, happy meat was better than sad, but marbled meat is delicious, so why not a month of corn in a factory barn? As to the killing, they'd rather not.

The next day Carl made a pot of chicken soup. We invited friends and family to join us. For about a week, there seemed to be chicken fat, chicken feathers, and chicken blood everywhere. I recovered from that relative trauma.

Though not really. Not ever. It left a gash in my soul remarkably like my long-ago Spanish adventure, a short-term experience that was more fulsomely developed during four years at boarding school. What of that cold night in hell? Would it, too, be a lesson I would have to learn again and in a more encompassing manner? Consequences, choices, and preferences.

Three chickens dead. Seven left. The next morning, I went out to the coop. The hens were much calmer, though Beatrice wandered the runs and coop aimlessly for days, squawking and running away when attacked by the other hens. She, too, was never the same. Her protectors and buddies had disappeared. They couldn't save her anymore. With them had gone her spunk.

Thus, life went on after death. Natural selection. Survival of the fittest. At Darwin's View. In retrospect, I realized that we had killed the boys on February 12, 2014, Charles Darwin's 205th birthday.

Beatrice

View of the top of Monadnock through a telescope

III
2014–2015: RESISTANCE TO RESILIENCE

To create pasture for sheep and other livestock, early settlers burned the forests of Mount Monadnock in and around 1800. A decade or so later, whether purposeful or accidental, human-wrought or natural, another fire ravaged the mountain. Orange and red flames outlined her profile as heat and smoke filled the air. The decay and detritus of the forest fed the inferno. It burned for weeks, leaving the corpses of scorched flora and fauna and fabulous views. People flocked to see them, climbing the craggy landscape.

Decades passed. Two centuries of regrowth. Grasses turned to shrubbery, trees, and the vistas of the nineteenth and twentieth century have been lost, enclosed by a new forest.

But not on the bald, granite mountaintop. Without tree roots to hold the soil, only wind and precipitation, erosion has left the summit bare. And then the persistent footfalls of humanity. Even so, a few shrubs and mosses have put down roots, leaving one to wonder if the summit view, too, will someday disappear, concealed by a canopy of trees.

Carl holding Cheeks

NATURE IS NOT MORAL OR FAIR

Chickens are uncomfortable creatures. They have rules we have no clue about. I can only put my impressions and imaginations on them, create a story of why something happened. For example, our Ameraucana chickens. Terrified and her sister Neurotic had never been accepted by the flock, maybe because, with their thick-feathered necks and muffed cheeks, they looked different. Neurotic lived a too-short life that ended, I hoped, swiftly, when that hawk swooped down on her. A brief squawk as the raptor grasped her in its claws.

That left Terrified to be renamed The Lonely One, then Lo, and eventually Cheeks. She grew up to look and sound remarkably like a crow. Of course, I knew she wasn't a crow, in part because she wasn't creative enough to jury-rig tools. In part because, when she reached adolescence, she began to lay sky-blue eggs. Sparingly. Her life was too stressed to focus on egg-laying. Though she had briefly been a favorite of Big Red, who as briefly protected her, too soon she found her place at the bottom of the pecking order. A not unfamiliar place to her. She and her sister had used to keep company there. Constantly chased and harassed by the other hens,

they had pressed themselves into a corner of the coop, or underneath anything that might protect them from the sharp rap of a beak. But now Cheeks was alone.

Through the winter, I would go out for brief stints to keep her company. She spent most of her time in the coop while the others explored, dust-bathed, and got their daily dosage of fresh air and vitamin D.

Cheeks

Cheeks ventured outside only if I were around. She would peck about near me but not too near. I would tidy the coop and runs until there'd be a squawk and a great flapping of feathers and she would be in my arms or on my shoulder, my cheek lightly brushing against her purple-sheened black feathers. At those times, I would look to Big Red and feel a bond. He knew. There is nothing like serving as a hen's protector to make one feel significant and needed. Eventually, though, I would have to extract her from my hair, detach her claws from my clothes. She clung, protested with guttural pleas as I set her down on the ground. This was a daily fledging that neither she nor I enjoyed. But as a chicken, she had to stay with her ilk, didn't she? It was for her own good, wasn't it?

I wondered at times. Did Cheeks miss her sister? Was that why she so vocally crowed at times, out of an unbearable loneliness? I worried about her isolation and lack of chicken companionship. Why had the girls ostensibly banished her? Was she unlikable? Were they jealous of her blue eggs? Perhaps Panda felt broody but didn't want that Poe-esque raven-looking bird around? Whatever the case, they killed her. It takes only one well-placed peck.

I assume they killed her. I went to check for eggs and there she was, dead underneath the coop, a translucent bubble coming out of her open beak, her claws wretchedly curled, and all I could do was think back to the morning, vaguely remember the sounds of the hens cackling, then a brief, chilling-in-retrospect calling out. I had thought it was in celebration of another egg.

Cheeks' death gave me a new perspective on chickens. I still liked them but I didn't trust them anymore. Their rules expressed nature's code, in which my definition of right and wrong didn't hold a place. Their social etiquette steamrolled over my meddling attempts at intervention. I had deluded myself into thinking I could make a difference, that somehow I could give Cheeks a happy, anthropomorphized life. How human of me! Because chicken rules and nature's rules: not for the first time, not for the last, I thought I could defy them.

As a child, I would sometimes get off the yellow school bus at the Art Yard, the studio and foundry my mother had built with two artist friends.

Mom in her studio: The Dome

I would step off the bus and walk down the winding dirt driveway to the open field, where stood two half moons made up of triangles forming pentangles, all connected according to Buckminster Fuller's design. Energy-efficient and wind-resistant, the geodesic domes sat like spaceships in the midst of the field, surrounded by open space, blue skies, fresh air. It was a piece of heaven seemingly created for Mom's beloved greyhound Géza. White with large markings of tan and brown splashed on his back and a mask over his face, he leapt and bounded through the tall grasses, ecstatic with the joy of running. When she called for him, he would return to Mom, panting and exhausted. She would pat her shoulders and up he would go on his muscular hind legs to rest his bony front feet around her neck. Face to face, they would smile at each other, she rubbing his soft, tawny ears, massaging his body. He would look about him luxuriously, as if to say, "Look! My perfect mistress loves me."

Back on all fours, his tongue hanging out, eyes bright, ears alert, he would trot behind her as she withdrew into the dome that was her studio and sacred space.

Inside: the smell of plaster and metal castings, clay, and melted wax. Gray cement floors, swept clean every day, exposed the tell-tale signs of

125

creation: drops and drips of dried clay, white plaster, paint. A pegboard wall held sundry drills, screwdrivers, hammers, polishers and buffers, clamps, and saws. A rickety metal bookcase full of paint jars, glues, and silicon sprays. Wire armatures waiting for clay. And two heavy, long, rectangular work tables, sometimes clear, more often covered with clay models and molds, works in process and finished pieces, all furthering her practice of the lost-wax casting process. Drawings and prints hung off the curved walls, and sculptures peopled the whole. In that dome, the air circulating, the radio playing classical music or NPR talk shows and news, she created, her youthful arms strong from wedging clay and then molding it into life-sized figures cast in bronze, aluminum. Her work embodied her self. Elegant and raw, primal: a trio of headless female bodies, one with claws for feet, another a torso slashed, an exposed vulva. A totem of two stunted bodies, faceless and turning. A diving figure. A striding figure. A rising figure. Distorted heads and writhing torsos. Discomfiting, the sculptures represented the awful beauty and piercing vitality of life. At times, its claustrophobic constraints.

My father and mother had attended college prior to marrying—Columbia and Barnard, respectively—and lived their first married years in New York City. There weren't a lot of jobs for East European historians at that time, so when Dad got the job at Michigan State University, he was happy for it, and Mom accompanied him. They found a lovely home out in the country with nice neighbors, if with different values: East Coast liberals meet the Bible Belt. Dad's work, and soon enough, Mom's, connected them to the art and intellectual circles of the area. They lived there for nearly thirty years. Exiled from the cultural and intellectual center of their youth, my father, an introvert, had his books. My mother, an extrovert . . . I imagine her sense of exile might have been exacerbated by the roiling feminism of the sixties conflicting with past expectations. Raised in a certain class and era, she had "come out" as a debutante and proceeded to live her life trying to get away from everything that class and era stood for. Ironic, isn't it? How sometimes we fight against something and the harder we fight the more entrenched it becomes without our even seeing it.

A flash memory, like a floater in the eye: I am around nine and sitting in the backseat of our red Chevrolet station wagon. My mother—her black, shoulder-length hair down, likely as not air dried, no lipstick or makeup necessary, buxom in a black, turtleneck sweater—is driving. She is stressed because she lives on the assumption that my father and she are equals. He works and so does she. But the childcare falls to her. And she is charged with entertaining my father's colleague while my father prepares for a class. Thus, my mother drives my father's colleague, a solidly built woman with a face that is round, flabby. Short brown hair. Her mouth turned down. Thick, round glasses that enlarge her rheumy eyes. She studies my mother as we drive out of our driveway onto Coleman Road, and then says, "You certainly are the perfect fifties' housewife."

Mom claims she had no idea at the time what the woman was talking about, though it was most certainly not a compliment. Both women were strong, intelligent, self-willed. The only difference: Mom was married and had two children.

They lived at a time of confusion that still exists in this society, in which success is based on work that earns money, preferably lots of money, and/or youth and beauty. In which case, what is a woman's role? What is an artist's? Is art worthy in a society that values earned money over expression? And who determines whose art is valuable? What of all the artists undiscovered? And all those who are discovered after death? What if a person's art is awful? Does that make their life a waste? And back to motherhood. Unpaid and endless, undervalued if at all, like the squandered wealth of nature. If one has children but has, too, other things to do, who is to say if that's right or wrong, and how dare anyone judge another's life, not knowing the choices and sacrifices made in its formation?

In her mid-fifties, my mother sold her studio and foundry and returned to New York City. My father continued to teach but would travel between East Lansing and New York City for holidays. My mother, meantime, eventually bought a studio in the NoHo neighborhood of lower Manhattan and began to take molds of the faces and bodies of friends, family, strangers. She cast living masks, made them smaller and smaller, distorting them as they shrank, still using the lost-wax casting process. Roman series. Estrucan series. Grids and abacuses. Books and walls of faces. Congregations.

Once, when I visited her early on in her repatriation to New York, she and I were late and racing to a show or museum. We ran up the hill from my parents' apartment on Riverside Drive, lifting a hand to hail a cab. In we got. She leaned back in the cab, breathless. She removed the soft gray mohair beret from her head, her black hair now cut short with hints of white. Her green eyes shone bright with curiosity and anticipation of what was to come. Finally, she was living the life she had needed. Feminine and feminist, she refused to return East Lansing. New York City matched her mood and soul: vital, bracing, crisp, and palpitating as a fall day.

"I love this place," she announced.

Then age. Parkinson's.

My mother started out an English major in college. She moved on to painting, then sculpture, earning her MFA at Michigan State University. She tells the story of her first casting. Her teacher—a man—pointed to her and the only other woman in the art class, stunning the men by allowing "girls" to do the pour. Empowered, they each took a side of the pole that held the crucible full of melted metal. They tipped it to pour the contents into the plaster mold. She fell in love with the molten lead flowing like quicksilver. The fumes wafted about her, caressing, titillating. Her body's energy joining with aluminum, steel and oxygen bonding—years later corrupting and no reset button.

Exposure to heavy metals can cause Parkinson's disease. My mother's passion in life, her work, arguably caused her disease. The consequence of that illness and its medications? Mental fogginess. Extreme exhaustion. An inability to work. Aware of her losses, she progressed from denial to anger to . . . where was she now?

My mother lived a dynamic, creative life. Given the choice—health or creation—would she give up having created? What makes life worth living? What is a worthy life?

Work comes first. I learned this form of prioritizing at my father's knee. A tall, lanky polyglot with riotous eyebrows, he also sported a sometimes-unruly mustache that would grow into long handlebars that he'd twist and comb but basically leave to do what it would, along with an aquiline nose that had been broken as a child. He was frequently thought to be of Slavic, rather than of Scottish, descent. His wiry, reddish-brown hair was receding by the time I came along and had an Einsteinian, distracted-professor look. His two hearing aids ever squeaked and carried on while he fiddled with the levels and batteries. In my mind's eye, he is dressed in a cerulean Oxford button-down shirt and khaki slacks. His hands, with their long, bony fingers, are held out to help shape and articulate his thoughts. A wry smile. A sparkle of amusement in his eyes. A wristwatch with a cloth band to keep the time.

Dad, years ago

He was a creature of habit. His dress code during the week consisted of Oxford shirts and khaki or corduroy pants. And, if he were home and not heading into town to teach, bare feet. He would breakfast—two fried eggs, a Thomas's English muffin with strawberry jam, bacon, and coffee—while reading the previous day's *New York Times*. By eight o'clock, he would descend into his basement study via a metal, single-helix spiral staircase, bending down to avoid hitting the scarred top of his head. The stairs would creak with every step. The floor of his office, a white-and-black linoleum tile, held the chill of the ground. The four cement walls supported shelves of books, hundreds and hundreds of books, that made up a library on European history and the concomitant socio-political havoc. The books added depth to the smell of must and mold of that basement. His study felt like a cave, isolating and dim as he sat down at his desk—a repurposed door covered with piles of books and papers centered around his green manual typewriter. Setting down his cobalt blue espresso cup full of strong black coffee, he would turn on a light and turn off his hearing aids.

I would sometimes sit at the top of those stairs and smell the dank air of the basement rising up. I shivered, and wasn't my father chilled, too, as he clacked out thoughts and connections on that typewriter? He appeared not to care. He was surrounded by books and ideas and utter silence. I was the last thing on his mind.

129

Once, in my twenties, I was on the phone with my father and asked him, off the cuff, if they had missed me when they sent me to boarding school. He replied that it had been nice to have their time to themselves again. He saw nothing wrong with the admission that I was, if not in the way, a distraction. Neither did I. Though I admit to feeling a twinge when he told me, my vague suspicion of insignificance validated.

Years later, when we sold the family's ancestral home in Stonington, my sister and I gave away and tossed out decades of stuff collected. The local second-hand shop celebrated the donation of kitchen paraphernalia, clothes, and usable odds and ends. The local animal shelters happily took on the old sheets and towels, and what they wouldn't take was accepted by the twenty-yard dumpster in the driveway That dumpster took in, too, my college notes from history, political science, and literature classes, along with broken shelving and furniture, moldy posters, and rusted-out pots. Friends would stop by and walk away with old paintings, books, vases. We let go of so much privileged stuff. Dividing between us, rejecting, we were exhausted emotionally and physically, at which point we found our father's boxes of research, boxes full of his last three years of work on the handicapped in Eastern Europe.

When he had died nineteen years earlier, we had called around to his alma maters, fellow historians, friends. His library of books, we donated as per his wishes. But his research notes? We could find no one to take them. And so, in 2012, with our mother entering the middle stages of Parkinson's disease, the decision was left to Xanda and me what to do with those boxes. We stared at them. Our mother had gone back to New York City. She was fully onboard with the sale of the Stonington house but was easily confused and made anxious. We couldn't ask her advice. We knew no one would use them. Pages and pages of my father's script, signifying hours and hours of research, of studying Turkish and Russian so as to be able to read the archives in Istanbul, travels to Turkey and Hungary and the Soviet Union, and thinking, thinking of the connections and themes and possible stories to tell, as historians do: through interpretation and analysis. Telling stories, shaded by opinion and colored by wit. Boxes of my father's mortal time—time spent with books, not people, not me—had developed a case of mold, disintegration—and was that time wasted? What matters in life? What makes a difference? All that energy expended, and had it been for naught?

Briefly, I thought I could do it: decipher his handwriting, learn the languages, fulfill his life. Very briefly, but then reality set in. My sister and I agreed. They had to go, those boxes, and I can still feel the gut-roiling thump in the pit of my stomach, the twist in my heart as we tossed them into the dumpster and his papers scattered in with all the other stuff of a lifetime, generations of what had used to seem important, and then? Now?

The world spiraling, rocketing through space. Species' extinction every day. Drinkable water more and more scarce. The season changes dramatic and the weather extreme. Mother Nature and mother in a slow death spiral and nothing I could do.

I began to wonder if I were writing my life into inevitability. I was a character in a book and it had to end. Happy or sad? Lost or found? I plotted a life at Darwin's View, ever moving, never pausing, seeking meaning even in the antics of chickens. Not just who was I, but why?

Carl was right. I had lost my sense of humor. I'd had it with change and nature's code and creativity. Mother Nature didn't care, either. She carried on. So did my mother, who had tools I didn't have, a perspective on life. My mother, who had gone back to Key West for another spring sojourn. Once again, Xanda flew down with her. Carl and I would bring her back up. A daytime companion while there, in case she needed help. Denise. I have a picture of Mom and Denise walking along a sidewalk. Similar builds and height, differentiated only by decades, genes, and haircuts: Mom's hair is practically shorn against her head, whereas Denise's black tresses curl down past her shoulders. There is a giant chess board in the background. They lean into each other, as if sharing a secret, connected by a *joie de vivre* and perspective on life that I was not part of but was grateful for, because, once upon a time, my mother looked into Denise's eyes.

"Are you happy?" she asked with a bright glint in her eyes that yet held her jaunty spirit, her humor. Denise smiled back, with her white-toothed, laidback Key West grin.

"I'm happy," Denise said, and her thick black hair bounced with enthusiasm. "Are you happy?"

"Yes! I'm happy!" They laughed, in fact and deed happy, proving Milton right again. The mind can *Make a heav'n of hell, A hell of heav'n.*

PROJECT PREPARATION

For all those warnings I'd received that they would inevitably be killed in some horrific fashion, no one had suggested preventing the chickens from staring into my office. That encouraged me to stare back. To note what tiny heads they had, and wonder, how big could their brains be? To recognize their personalities and attachment to each other. To look deep into Big Red's orange eyes, watch his head turn and twist nearly upside down as he tried to see exactly what I was up to. In the end, I had gotten kind of a crush on that big-hearted Rhode Island Red, who was initially supposed to be our main layer and he was, just not in the way we'd anticipated, and there he stood, in all his cockiness, and what were we going to do? Now I'd gone and gotten used to him. He was practically family.

Bustling and busy, he would greet me with a lowered wing. A regular Fred Astaire he was, dancing and chortling, his activity simultaneously stressful and reassuring because his "inevitable" and "horrific" death was still in the offing. Not so charming was the chicken manure. It had to be dealt with every day, and took on an extra air and urgency in the midst of summer's heat and the chaos wrought by a broody hen.

Yes, life counters death, and right on schedule, Panda went broody just as we left for my family summer reunion. Hunkering down with fluffed determination, her usual mellow cluckings deepened into a throaty croaking for chicks and a refusal to eat or drink or scratch. She plucked her chest bare so as to allow her body heat to warm the chicks when, if ever, I gave in, and it was my bad that I had not yet provided her with newborns. She was still in that state upon our return and I responded in kind. I searched unsuccessfully for locally hatched chicks and thus ended up mail-ordering—against all promises to myself not to, but how reason with a broody hen? I apparently can't say no to one. Four chicks. Golden Buff Red Stars. A sex-linked breed, guaranteed to be girls because the boys are white when hatched, the girls brown. No mistakes possible.

Actually, three Red Star girls; the fourth was an irresistible Lavender Orpington chick who was not destined to be my best buddy. Opie arrived ill-formed and doomed, determining me never, ever to buy a mail-order chick again. Never.

We buried Opie with Cheeks and Chipper and left the Golden Girls to grow up and enter life's pecking order. They scurried about, hovered over by Panda and chased by Beatrice, who appeared to be jealous of Panda's attentions, confronting her mother for her desertion, going chest to chest. Beatrice in her lonely rebellion, who had landed smack dab at the bottom of the pecking order when Cheeks died.

Beatrice was a small hen, reminiscent of Lola and as spunky, determined to make one of those three blonde chicks—Brownie, Clownie, and Downie—fall below her. The other hens kept to the outskirts, out of Panda's protective way, eyeing the chicks as they grew into pullets. Big Red had his eye on them, too. He and Panda tussled, she having entered her bored-with-motherhood stage. She wanted back into the flock and her place at the top with Big Red. Cutting to the chase, Panda introduced the adolescents to the big coop. Typical of youth, they hopped onto the top roosting pole, thereby usurping all the best seats. A hubbub amongst the hens ensued, but the young pullets had Panda to protect them and Big Red knew better than I not to interfere.

Meanwhile, our "right-sized" house felt small and claustrophobic. There was nowhere I could stretch without whacking my hands on a beam or a piece of furniture. And in timber frame houses, sound travels along the beams. My office didn't feel much like a room of its own, given Carl and I could carry on a conversation, I in my office, he in the dining room, with doors closed. To distract us from this luxury quandary, we decided to renovate the chickens' quarters by moving the nesting boxes and roosts. It would create more room in the big coop, now the flock was up to ten. And a poop board would allow us to remove the chickens' excretions easily and effectively. Add hen hubbub to the pecking-order pandemonium and we had ourselves a project. Carl went outside to negotiate with Big Red.

Construction work tended to bring Carl and Big Red to fisticuffs. Big Red considered anything "outside" as his territory. That included all things hens and coops, and coops included structures of all sizes, not least the house. Thus, anytime Carl went outside with hammers and Dewalt electric drill, Big Red came at him, spurs first. Until, one day, Carl put down his drill, ready to take out that Rhode Island Red rooster . . . and

Big Red set his sights on that dastardly yellow-and-black drill. Ever since, they had an understanding. Carl would march outside to set up whatever job was to be done. Big Red would approach warily. Carl would put his electric drill on the ground. Big Red would attack it until it was good and dead. Then Carl could carry on with whatever he had to do, leaving Big Red to crow his disdain and strut his mightiness from a distance.

March out. Approach warily. Drill on the ground. Annihilation. With that process wrapped up for the current renovation, Carl proceeded to remove the nesting boxes and create a new roosting situation. I cleaned out the used litter—our future compost—and replaced it with fresh-smelling pine chips and hay. A dusting of Layer Herbal blend to calm ruffled feathers. Carl and I admired our work, then opened the door to allow the chickens to pass their judgment.

First, Big Red and Panda, with the babies in tow, studied the situation, side by side, pacing the coop, heckling and clucking. Big Red made a running commentary in his low guttural tones, pointing out the new cozy spots with excited cluckings. The little ones tested the height of the roost, Panda the walk into the nursery where the nesting boxes were located. Enter the other hens. Bafflement and much wandering about. Pecking and squawking. Beatrice ran the gauntlet in a panic of where she would end up. There would be a new order; every evening I'd find different pairings on the roost when I went out to put them to bed. It was a chicken soap opera. Who got the prime seats next to Big Red? Who was roosting next to whom on which perch? Some nights, the hubbub was too stressful. I couldn't watch. My attempts to help only made things worse. I would force myself to walk away, leaving Beatrice crouched under Big Red's chest. The bigger hens pecked at her but she would not give up her place. The littles shuffled and pushed themselves under Panda, motivated to not be where the other hens could peck them.

Raising chicks, if you're insane enough to do so, is much easier with a broody hen to manage the precarious rocketing about of the peeping balls of fluff, but there is still the relentless worry about the flock that can inhabit some brains, specifically mine. Did they have enough water and food? Had the humans remembered to lock the coop door? Was Panda on top of the pasty butt situation? However, would I have chickens that weren't feral if Panda insisted on treating me like a pariah, thereby causing

Learning to fly!!

the chicks to run from me in terror? At which point the worry morphed into more piercing questions, such as if there were a hurricane or fire, how would I save them? They'd run away from me, thus die. Would that be a relief and blessing? And if I had had any idea how much brain space, dream space, money, stress, and time the chickens would take up, would I have adopted them?

Absolutely! When we learn life's lessons, they are earned at too great a cost to give them up. Like boarding school. I am who I am as a result of that experience. Just so, I would adopt chickens again because they have formed me, in their odd and quirky way, and directed my path, probably more than has any bird in human history.

Granted, by then I had to admit that it was not Big Red and the hens that kept me at Darwin's View. That was an excuse. At some level, I wanted to be there. I was working my way toward that idea: that for all my regret of the past and dread of the future, the present was where I needed to be. But what would I do there?

Construction. Destruction. Deconstruction. Was I merely a chicken shepherd? A humbling thought, especially when visiting my mother and sister in New York City where life revolved around art, theater, their own work, and creativity. Whereas Xanda would expound on the most recent exhibit she had seen, play she had attended, or review she had read, my additions to the conversation consisted of updates on the chickens and coop renovations. A country bumpkin trying to fit in in New York City, and at Darwin's View, a city gal attempting to be country.

WHEN IS IT MY TURN?

"When is it your turn?" my mother asked me. The heartfelt question was full of my heartbreak, because what could I say? She lived in New York City and I between two places, but more and more in Jaffrey. I visited her as often as I could, though not as often as I needed. Our relationship resembled the Star Trek transporter when it wasn't working and the person beaming up faded in and out, and was sometimes lost forever. I reached to grasp her. With that question, did she reach for me? My turn to what?

She couldn't say. The words disappeared. My mother's increasing inability to communicate was like writer's block. Beautiful ideas are perfectly formed in the mind, but as they approach the tip of the tongue, when pen is placed on paper, the words and expressions dissipate, leaving a sense of loss and disappointment that grows with every next attempt. Days and weeks become months and years of foggy thinking, unarticulated thoughts and feelings, with an occasional llama or fluffy-tailed cat for entertainment. Hallucinations—what we called creative visualizations—that amuse for a moment, then frustrate and depress. The bugs. Wasps. How is it no one sees these white pieces of paper falling from the sky, filling the air? My mother's querulous questions and it's pill time. Again.

Years ago, one of my cousins had twins. Those two pips are men now; I have no idea how that happened but it did. One of them studies black hole spins (astrophysics). The other studies cell structure (biochemistry). One egg split in two, and we have twins who study the biggest and the tiniest aspects of the universe, in a yin/yang way because they explore the same thing, if in different forms: energy.

Energy is complicated to describe if only because everything in the universe is constructed of it. Fortunately, I go for the big picture, not the details—remember, when I moved to Darwin's View, I told the energy consultants that I wanted the light switch to work and don't care about the how of it; that's Carl's job and Mother Nature's—so I think of energy as fuzzy balls of movement, electrons that don't exist in a physical sense because movement only exists relatively (see the Theory of General Relativity). So if you try to look at an electron by taking a picture of it, it isn't there because it is movement, and even by attempting to photograph it, you disturb and, therefore, change it (see Heisenberg's Uncertainty Principle). Electrons hopping around the universe in various forms. They jump from one ring to another, bonding together to make minerals and rocks or the atoms that make cells that make us. They careen back and forth between me and my chair, Carl and me. The chickens and cats. Energy connects me to everything around me, even to some people I'd rather not be connected to, but still, it's an incredible thought to be so much a part of something and ever-changing. Like a shimmering web of vibrations. Go to the very edge of you, and you will find you are a part of someone or something else.

I get dazzled by that fact. Microbes and mycelium. Plants and animals. The soil and stones. Water and fire. The stars and planets. Mothers and mother trees. All energy. All movement, and movement is relative. Depending on how one thing moves, so does another. So simple, yet overwhelming, because if we are all connected in ways we cannot see but are minutely and directly affected by, then it matters how we move and what we do. Everything we do matters.

This logic added a degree of pressure to my actions. It meant the world, maybe even the universe, depended on me.

At least, the chickens did. Not so much my mother. How often over the years had I gone to visit her in New York City, thinking I'd have time with her, and arrived to find her in the middle of a meeting or her day already planned. I was welcome to join her, of course. Her offer to live her life with her, not mine, whatever that might be, translated into a sense of rejection. Always her schedule. Her own deep need to live. Alive in New York, she breathed in museums, operas, art exhibits. She had a cultural, artistic, human existence, and diverse friends who traveled, communicated, created exactly as she did. She was gifted ten years, more or less, before the disease took over.

Once upon a time, near the beginning of her tango with Parkinson's disease, I left my mother in her New York apartment to return to Providence. So much easier said than done. I kissed her good-bye and headed for the door, but then turned back for one last kiss, and out the door, where I stood in the entry, waiting for the elevator, debating . . . but the elevator door clanged open. Greetings and comments on the weather with the doorman. Down we headed, then out of the building. I walked across town to Sixth Avenue, then Seventh Avenue, because I needed to get home to Providence, to Carl and the cats, my commitments and friends. But leaving my mother felt wrong, as if I were deserting her. Didn't she need me? Her eyes had held a distant loneliness as I kissed her good-bye, and the apartment would be so quiet. Hardly bearable to think that I would inflict on my mother the piercing silence of isolation, the heartbreak of absence with which I was so familiar. By staying, I might alleviate that desolation. That thought caught in my throat. I might matter. I would make a difference. I had to stay and so turned around and headed back to her apartment. But then Carl. The cats. My life. I felt their pull, their needs and my need for them opposing my mother's. I turned back uptown toward Pennsylvania Station. Walked a few steps. Tears welled as I walked. Surrounded by the honking and bustle of New York City, my heart squeezed. I couldn't get a breath and turned downtown again. She needed me. I stopped. So did Carl. What did I need, want? What should I? Trapped in indecision, neither choice right, I called my mother. I wept as I used to when I'd call home from boarding school. I asked her what I should do. I didn't know what to do. I would stay if she needed me. A pause.

"Someday," she said, because then I still needed her more than she needed me, "I will need you to stay, Tory. But not today."

Tick. Tick. Time passed and I, the suspect philistine and proven country mouse, now tidied the chicken coop and fantasized about creating a place to which people might come to heal. But not my mother. Just as I would not move to the city, so she would not move to the country as the baton passed, mother to daughter, and who takes care of whom? Communication grew worse. She became less and less able to articulate her thoughts. We did the best that we could, and it was often imperfect.

Imperfect was not enough.

It had to be, because a lowering baseline didn't change the essence of what was. Mother Nature. My mother. My primal recognition of the vast pool of absence their leave-taking creates.

I needed her then. I need her now. Her absence fills me. And so when my mother asked me, "When is it your turn?" I comforted myself by

138

thinking how we are connected at a cellular level, energetically. We are one, if divided. And when my mother asked, "When will you be here?" I think, "*Every day, every hour,* in spiritu."

That summer, during our annual family reunion, Xanda and I realized that our mother was far more compromised than we had realized. Denise, from Key West, had come up to assist over that summer holiday and heard us debate the logistics of hiring a full-time companion, one who would live with our mother in New York and travel with her. Denise said she'd be interested in the job. We hired her. She moved up to New York City that fall. Imagine that: a Key West twenty-something moving to New York City under the tutelage of our mother. A place to live. Opportunities abound, and how grasp them?

At that same age, I'd lived in New York City. I lasted six months, working at a publishing company as receptionist/office manager. I tried very hard to fit the bill. But, at my first review, the woman who had hired

Denise and Mom

me and I agreed it was not a perfect match. The publishing firm found an organized office manager, and I, after attempting a freelance writing job— that I eventually recognized to be writing college students' papers for them, and so quit—returned to Providence and a job at a bookstore. All to say, Denise had a derring-do that I did not have. And that when Mom visited, she now had Denise with her plus nighttime aides, and life got crowded with my mother around.

I didn't care. I wanted her with me. Then I might help. I might save her. Thus, her visits mattered. Deeply. It was the only way that I could be with her and also with Carl and the cats and chickens. Home. I knew travel was not as easy for her as it had once been. She preferred New York City to Mud Mountain and the terrifying Big Red. When she visited, I wanted her to be happy, comfortable. Meals became marathons that Carl and I had little training for, being more *ad hoc* in our dining. My typically ovo-lacto vegetarian ice box would fill up with bacon and lamb, steaks, hamburger, and, yes, chicken. Gallons of ice cream, cookies and chocolates. Angel hair pasta was preferred over fettuccine, and no mashed potatoes or butternut squash. Mornings, I dragged myself away from her, ostensibly to write. I would listen to Carl and Denise entertain, and feel an absence and longing to be with them. I would hope I would be called for, needed. Eventually, I would make some excuse to leave my office and join the conversation. And then it would be time to drive Mom and Denise to the Bond Wellness Center for exercise and physical therapy.

Every day, preferably, we would go to the Wellness Center, because Parkinson's demands constant use of the body. Move those muscles or they will tighten and atrophy. The physical therapist there was a phenomenon. A petite woman, she expended more positive energy in one hour than I could in a year. Just being with her energized a person. One felt more hopeful. In a week, Bernadette had my mother walking, not shuffling. Stairs still challenged her, but she went up them, if breathlessly. Her voice sounded stronger and her mind was clearer. She was more awake and alert after her workouts. I fantasized about having Mom live with us, working with Bernadette every day. She would regain all she had lost, get back to her artwork, converse, be my ever-moving, active mother again.

Bernadette had a day off. My mother was tired and wanted to stay at the house, but Denise and I encouraged her to put on her winter clothes and baby step over the ice and snow to the car. Exercise was good! She would feel more awake and energetic! We arrived at the center and my mother refused to get on a bike. I suggested she walk around the track with me. She argued. She wasn't up for it. She was tired. I suggested the royal we try. We walked, slowly, back to a shuffle. She panted with the exertion, leaning heavily on my arm, a sheen of sweat across her face and her eyes wide and staring.

"Heel, toe, Mom. Heel, toe or you'll trip."

After one lap—one-sixteenth of a mile—Mom said she wanted to leave.

"We just got here, Mom."

My only excuse for what came next, as she took a few more steps and then came to a stop again, is that the situation, writ large, hit a button. Big, red, and ignorable because it concerned the heart. My heart expanding, contracting, bleeding with the panic and suspicion that all the progress she made with Bernadette would disappear when my mother returned to New York City. We had wonderful physical therapists and assistants working with Mom, helping to make her life easier, getting her to the doctors, her therapies, but life took over. The days passed. Exercise, from my distance, seemed not to be a priority. She preferred to go to her studio, rather than out to Brooklyn for a Mark Morris dance class. Staring at her where she stood in the middle of the track of the Wellness Center, depths welled, of frustration and anger and sadness that this was it. If my mother didn't overcome her physical limitations in New Hampshire, she wouldn't. Ever.

That thought caused my heart to palpitate and my brain to click in, overriding feeling. Fact: we had driven twenty minutes to the Wellness Center. It was ridiculous not to stay and try a bit harder. She had to make more effort! She acted as if on a whim. As if she thought she were the exception to the Parkinsonian rule that one must constantly retrain the body. Muscle memory was short and atrophy not far behind. Mind over matter! Ignore the tiredness! Walk!

Her eyelids drooped with exhaustion.

"Mom, you have to do a little more."

A tussle of wills and I said in a too-sharp voice, "If you don't work harder, you'll end up in a wheelchair."

We both fell silent, I mortified, and my mother? Exhausted by the disease and her relentless struggle against it. I apologized and motioned to Denise, on a StairMaster, that we were leaving.

It was a self-inflicted wound that took me days to forgive. I had been cruel. I had pointed out an inevitability and, potentially, destroyed the only thing that Parkinson's patients have: hope.

SEEING WHAT WE WANT TO SEE, OR NOT

A frigid November, and my chicken obsession had reached new heights, if that were possible. I still perked up when I approached the coop and there they were, gathered expectantly at the door, anticipating me and the treats I offered. Upon closer inspection, however, those most spoiled of chickens had *issues*. An untimely molt in late fall. Feathers everywhere but on the hens as the temperatures dove into the twenties. The girls looked, well, I didn't look because they were basically naked. And none laying.

I didn't blame them. Without their downy feathers, it was too cold to lay an egg. The days were too short. I suggested to them that they focus their energy on growing back their pin feathers, which they did do. Their feathers were reassuringly filling out, right around the same time as Big Red came down with his annual case of frostbite. It appeared daily to be worse, his comb and wattles swollen and black and falling off. Thanksgiving approached. Ping developed a swollen eye. Our vet told us to separate her from the others, put her on antibiotics and—very important!—contact him if any of them began to sneeze.

Separation anxiety ruled as we placed Ping in one of the runs, in sight of the others but not physically with them. Our chicken sitter arrived and we left for Thanksgiving holiday in New York City. When we returned, Daffodil had swollen eyes, CooLots and Clownie were sneezing—allergies?—and Beatrice had bumblefoot, a bacterial infection in her foot. A call to the vet. We freed Ping from her solitary confinement. She raced to rejoin the flock. A happy kerfuffle of clucking and coos, they not realizing what was about to happen. I approached the coop, paused to contemplate their obliviousness, their mortal purity. Then, with regret and disappointment, I put antibiotics into the water of our happy, organic chickens. We made note of the ten days they would be on them, and the two weeks after that until we could safely eat their eggs. If they ever laid an egg again.

Antibiotics might have saved the chickens, but not their few eggs that began to arrive, one or two a day. We tossed them over the wall facing

Mount Monadnock. Fourteen-year-old Peyton—her lower lip pierced by a round silver earring, her thick hair cut short and dyed various shades of deep purples and blues, her eyes deep, dark brown, and clearly seeing us as we might be interpreted as being because, once again, we were acting crazy. She pointed out, with an indulgent smile, that the eggs we were tossing out were no different from any eggs one might buy at a grocery store. I replied, perhaps a bit shortly, that that was exactly why I had bought into this chicken life: to have healthy eggs, ones not saturated with antibiotics and bad karma. I left it to Peyton if she opted to eat the eggs and headed off to the farmers' market to buy some I-hoped-they-were-happy eggs.

The nuclear clock, in the form of climate weirding, ticked to midnight, the calendar toward 2015. It was a brutal winter once it got started. Lots of snow all at once. Arctic cold. I wondered if the Gulf Stream had flipped. The ice age was here and we would never see spring. Thirty-five-mile-per-hour winds that lasted for hours, for days. The house shuddered. I reminded myself to be calm. Relax. Big Red crowed with whole-body passion, but there were no little birds in the skies. Would they come back? I missed the perky little birds hopping about outside my office. The plump tuxedoed phoebes puffing their chests. The iridescent blue barn swallows swooping. The kestrel hovering, its wings balancing on the passage of the wind. Snow buntings swirling, doves cooing. Still, our chickens pecked and cackled. They kept me brief company, though they brought to mind all the others that had flown, perished. The dying frog species. Butchered elephants and rhinos. The turkeys, cows, and pigs suffering in farm factories.

I girded myself. So did spring. Not quite ready for flowers, she presented us with mud. The chickens, freed from their runs, raced across the driveway, their wings flapping as they slipped and slithered along the path to the front door. Protected by a drift of snow from the wind, bare ground to peck at, they perkily stared into the porch. Did they really expect to come in?

Did I really not see the commonality? I was just like them, expecting the unthinkable to happen.

It does. Sometimes the unthinkable does happen.

My mother developed an eye carcinoma. This, apparently, occurs in some Parkinson's patients, possibly caused by the medications. She had to have an operation. The consequence: weakened eye muscles, thus frequent double vision. It was hard to tell what she could and could not see. "When is it your turn, Tory?" became "Where are you?"

"Here I am, Mom," I would say, waving to her over FaceTime or Skype, and she would look in my general direction. If she could hear me, I must be somewhere near. Though I wasn't. I was in Jaffrey, a five-hour trip one way if I drove, which I didn't. I took the train and so it became a seven-hour trip, door to door, and thereby required an overnight stay, or four. Spontaneous trips no longer happened. A computer screen had to do. I told myself to be grateful. At least I could see her. I could watch, fascinated by her movements, so like an infant's squirming, testing the extent of its world.

My mother was still a presence but fading. Shakier on her feet. Weaker and subject to the nausea of exhaustion. Her thoughts floated up, then, as quickly, disappeared, only partially articulated. She would struggle to bring forth a thought but her voice had no force behind it, and too often, she was unable to tie three words together, never mind describe the complications of a remembered event that might have happened. Then again, maybe it didn't. Memory, dream, or hallucination, it didn't matter. I wanted to understand her so that she wouldn't feel isolated, misunderstood.

Unthinkable. She was just like my father. A previously hearing deaf person, he had been desperate to connect, express, understand the world around him that once was and now is no longer. Sometimes, when a conversation reached gale proportions, swooshing from one topic to another, my father had used to pretend to hear and understand what was being said, "passing" the profundity of his deafness. Just so, my mother. I would look at her sitting in her chair, leaning at a tilt approximating that of the Tower of Pisa, her eyelids quivering as she fought to keep them open, stay awake. She made a good game of it, but was she processing it all from where she floated, on the edge of the subconscious and dreams? Was she as lost as I felt, trying to fit in?

At which thought she would look directly at me and articulate her response to a news story I had just read to her, or a reference to a conversation overheard the day before while she was resting in the other room; my mother has excellent hearing.

Those moments amazed me and horrified me, because they proved that her mind might be mired in the mud of disease but it still spun, active, determined, dying to live.

Like the sighting of a bobcat, those articulations were more and more rare. The fewer there were, the more I needed them. Which meant being together, not apart. I knew that, when she and I were together, the possibility existed for us to connect, and thereby I might exist at a level that had been buried by years of our mutual neglect. All those years when she was independent and free, where was I? The good old days, when there was a

Little Big Man, Peaches, and Apricot as peeps

slow-motion shift from August heat to September's chill, October's crisp air and shorter days changing leaves from greens to reds and browns. Fall followed by winter, then spring—and change is not easy but we do it as naturally as the seasons. And who was I, who was she, who were we then?

And now?

Passing. How many people do it? Pretending to be other than they are out of shame of being less than? We live in a society in which the expectation is "bigger is better," and "you can't be too rich or too thin." From personal experience, I knew it was possible to be too thin. I was learning that bigger won't be better when our singular world runs out of resources, which it was fast doing. And I knew that money holds a moral responsibility even if our society doesn't recognize that fact. But what did all that mean in my day-to-day life?

From the unthinkable to the unimaginable: that my fantasy of healing the earth might come true. How so if I couldn't protect Big Red's wattles from frost bite? Or my mother from a living death?

At least I could put on a Band-Aid, in the form of the third batch of chicks I ordered when Panda went broody, again, and once again, against all my promises to myself not to order from a hatchery. Leaving nothing to chance, this time I ordered Cream Legbar chicks. They are auto-sexed. Similar to sex-linked chicks, you can tell the boys from the girls when they hatch. Foolproof! One knows beyond doubt. It is guaranteed at a genetic level.

I ordered the four-chick minimum, ignoring their exorbitant price tag and including voluminous notes on the importance of sending me only female chicks. I would take anything, but not boys.

They arrived, cheeping adorableness, though one did look remarkably different from the other three. He, as he turned out to be, fortuitously died a couple of days later.

Three left, one of them not quite a girl. BackYardChickens experts agreed it might be a boy. I thought so, too. But MyPetChicken said it could be a girl. We needed time. It ain't over until the fat lady sings, or the cockerel crows.

Sunflowers towering

NO GUARANTEES

Late summer of 2015. Sunflowers taller than I by a foot. Yellow and orange nasturtiums growing up black-flowered hollyhocks. A most satisfactory crop of garlic. Providence had become a way station for friends and family, rarely us. No eggs. For weeks, no one had laid one. We blamed Panda's broodiness and subsequent motherhood, assumed it had upset the other hens' game. Ooooh, they couldn't bring themselves to lay an egg. How could they, with the passion of Panda in their faces, showing them as lesser for not being the mothering types? Isn't mothering what hens are supposed to do besides scratching and cackling? Eventually, once Big Red and Panda began mimicking Carl and me by acting like an old married couple again, one or two eggs began to appear in the nesting boxes. Normal was the new norm. The girls were back to work. But as the babies grew up and moved into the main coop, Carl reached his limit on chicken manure and suggested that maybe it was past time to take out the poop board. Originally put in place to aid a chicken sitter in cleanup, now it served only to disgust us. Every morning, a large spatula to clear the board, a bucket to dump it into. Their crap might have been fine soil amendment, but it stunk and stuck and made for a distinctly unpleasant morning chore. We were ready to go back to our study of deep-litter manure management systems.

Carl got his drill, I got my iPhone, and we headed to Chicken Paradise. While Carl performed his tango with Big Red, I went into the nursery to take more photos of our pullet-who-might-be-a-cockerel. It was all so edgy, hope vying with the reality that we had received a fifty/fifty ratio, boys to girls, of *auto*-sexed chicks. I entered the nursery run and the chicks scattered, peeping in their alarmed manner. Panda raced about, fluffed, clucking her warnings. I rolled my eyes at her, because, really? Couldn't she cut me a break? I laid down on my stomach and crawled under the nursery to get closer to the feathered tots who proceeded to scoot past me, back out into the run. And then a loud clunk and a curse from Carl in the coop.

"What?" I shouted. "Are you okay?"

No response. Had Big Red won their *pas de deux*? Or had Carl succeeded in slicing his neck again, as he had the previous summer during a Providence bathroom renovation, when reaching up to pull a piece of Vitrolite glass off the wall? It broke and fell, neatly sliced into his neck, barely missing his carotid artery.

"Carl?" I backed out from underneath the nursery, imagining him dead in the coop, which was and is not the vision of sanitariness. I stood, whacking my head, and cursed, because, for all our moving about of the runs, we had yet to manage to make them high enough to stand in upright.

"Unbelievable!" said Carl, apparently alive.

"What?" I ran to him, hay and chicken dander clinging to my clothes. When I opened the coop door, expecting blood, there stood Carl, holding the poop board and staring down at where it had been. Staring, too, I admitted defeat.

Two months before, I had stuffed, tight as could be, three bales of hay under the poop board. I had shoved and pushed the hay and called Carl over to do the same for the very specific purpose of preventing the hens from going under the poop board, and for insulation from cold. Hens are not mice who have skulls that can squeak through holes as thread does a needle. There was no way a hen of any size could crawl or shove her way under that poop board. We thought. The girls, once again, proved us wrong with the proof being in the pudding. Or the hay: Two nests chock full of eggs. Lots of eggs. Seventy-six eggs to be exact, perhaps as a doff of a cap to Carl and his trombone.

Carl took over the egg removal task while I uploaded chick photos onto the BackYard Chickens website. Hiding the eggs was entirely understandable. Hens do not like having their eggs taken. Perhaps their distress isn't at the level of cows' when their calves are taken from them. The deafening uproar and bellowing of those unnecessary and heartbreaking separations, I hope never to inflict or endure. But hens do exhibit umbrage when they note my scavenging hand in the nesting boxes. I try to make sufficient oohs! and ahs! and so beautifuls! and thank yous! that they tolerate the thieving and return to their scratching and dust baths. Seeing seventy-six eggs, though, gave me pause. Did they actually care for each and every egg, or was it the principle of the taking? It has been shown, time and again, that when a cow meets again with its calf, now grown, they recognize each other. Why not an egg and a hen?

First of all, because an egg is only a potential chicken. Fertilized or not, it has not grown into anything to be recognized. Even so, might they not

Seventy-six eggs in their nests

hope? Might I not? I noted the distinct possibility that some of these eggs might have been warmed to the hot temperature of a broody hen's body. That some of these eggs might become chicks if given twenty-one days under a nice warm hen, or a heating lamp.

Carl responded to my question by filling a large, stainless steel tub with water and dunking all the eggs.

Egg shells, in the way that all matter is actually just a bunch of random energy bouncing about, are not as solid as they seem. They are, in fact, porous. As they age, the mass within shrinks and is replaced by air and gases. Thus, when an egg is put into a bowl of water, it will sink if it's fresh. (More mass.) And float if it is old. (Less mass, more air.) In other words, old eggs float, fresh eggs sink. Forty-five of the seventy-six eggs sank. We put the good eggs into cartons. Carl composted the thirty-one bobbers. I began to contemplate recipes for soufflés and crème caramels, along with dinner-party guest lists. The BackYardChickens poll came in: she is a he. I announced the news to Carl through the door of the closet, into which he had disappeared with a dozen of the eggs and a flashlight. He came out and, together in the dark of the guest room, we lit up the eggs with the flashlight, thereby candling them to see if they were fertilized and viable or merely eggs to be scrambled.

"They all look the same to me," I said.

"This one might have been the start of one." Carl, ever the optimist, pointed out what looked like blood vessels to him and egg shell to me.

"Will we be able to hatch them?" I asked, forgetting that, by getting them wet, the protective coating on the egg, the bloom, had likely been removed. An irrelevant point anyway, because Carl's conclusion and reply: time for breakfast. In the kitchen, he took an egg that had passed its dunking inspection and cracked it open. The smell of rotten egg permeated the kitchen. He tried another one with equal success.

Carl composted the rest of the seventy-six eggs. The hens cackled triumphantly in the background. A few weeks later, Little Big Man crowed his first crow.

For once, I didn't embrace denial. Not entirely. Because Carl didn't seem worried. He had, apparently, heard my justification for spending oodles of money on Cream Legbar chicks: we could start a breeding program, sell them. That idea, of course, neglected the tiny fact that I wouldn't let go of any chicks we hatched because s/he would be my responsibility. I would not sell a cockerel or pullet because then that sentient creature would be out of my purview to protect it, not from life—it was dawning on me that there wasn't much I could do about that—but from humans. That's how far my controlling nature reached. My possessiveness went beyond my kitchen. I wouldn't give up the eggs, never mind the chicks, to someone who might torture, maim, or kill them.

And so—however contrary to my past actions this might have seemed—I could not justify breeding chicks only to cast them out into the world, thereby enabling humans to add more bad karma to an already overload of CAFO bad karma. I held out hope that, in the midst of life's chaos, I might do less harm if I couldn't do no harm, and watched, resigned, as Little Big Man grew up. I imagined he would become Big Red's left wing, assisting in surveillance duties.

What comes first, the chicken or the egg? As any hen will tell you, what does it matter? You have a fabulous feathered dinosaur in one hand, and a stunning, perfect oval in the other. The more relevant question is, what's next? Treats, or tricks?

Big Red

ONCE UPON A TIME

Big Red molted around the same time the hens did. Just as the fall chill set in. I didn't know roosters molted. No more tail feathers, prim and shiny. His green-and-black feathers that used to be a boa around his neck, gone. He looked like a sad rendition of a chicken, but they all did. Even CooLots had shrunk, looking absurd with her feathered legs, and that was about it. Naked, but then hints of incoming pinfeathers. The difference? Whereas the hens continued to cluck and peck about, Big Red seemed . . . off. His crow kind of split into a dying off he hadn't the energy to finish. Rather than sleeping on the roost, he moved to a corner of the coop under the roost where the hens unbecomingly shat on him.

My chicken *aficionados* at BackYardChickens suggested that he was protecting his coop. Maybe something had threatened to come in, and he was bracing for the assault. I was so proud! Big Red was so brave! Such heart in a rooster. Who would have thunk? Because we had, indeed, just had a minor hiccup in the coop. The solar-powered door between the coop and northern run had stopped working. For a few nights, unbeknownst to us, the door had remained open and *anything* might have come in and wreaked havoc on our flock. Big Red, though, was on the job. Resting on the coop floor. Alert, if covered with manure.

A part of me questioned that assessment. I almost dared to think that Big Red had plummeted to the bottom rung. Was it the presence of Little Big Man? Hard to imagine. Little Big Man was a dandy with his floppy comb and pathetic ca-doodle. Big Red's was the original tune. But who was I to judge? Maybe Big Red, exhausted by the thought of that cocky youth, had decided to toss in the towel, let the little guy deal with the stress and strain of looking out for the flock. Big Red joined the gals, pecking and hanging about while Little Big Man practiced his tinny, out-of-tune crow,

Little Big Man

an entirely different song than Big Red's. Big Red's was stronger, louder, mightier. But, for all his ta-do, I knew Big Red was a pushover, even if everyone else held him at a distance, afraid. Not least Little Big Man, who was a skittery, hyper fellow. The girls were unconvinced by him. His comb flopped about. Big Red's flaked. More feathers fell out. I hoped it was just a heavy molt, but he certainly had aged and shrunk, ancient at three years and four months.

Panda was beside herself, approaching him, ignored. Not even the Golden Girls, crouching in front of him, could get him to perform his cock-a-doodle duties. As a distraction—whether for herself or me is beside the point—Panda went broody. In the middle of a very cold November, she bared her breast of feathers, fluffed and huffed, and sent the other girls running away. For once, I stood my ground. Every day, two or three times a day, I'd go out and remove her from her nest, carry her around, allowing her chest to cool. I read that I should put her in a wire dog cage balanced on two sawhorses for a couple of days. That seemed cruel. Instead, I continued to push her off the nest, distracted her with treats. I locked her out of the nesting area during the days, leaving her pacing next to Big Red, who commiserated with her from where he stood in the sun.

I successfully broke Panda of her broodiness. Part of me wished she had been more persistent. We still had a very sick Big Red on our hands.

Maybe she had hoped to cheer him by providing him with cheeping Rhoda Reds for Christmas. At this thought, the small part of me that didn't regret breaking her broodiness did a brief victory dance before returning to reality.

"Just in case," I said to Carl, as I fetched the cat carrier to take Big Red to the vet. "How awful would we feel if something happened to Big Red and we could have prevented it? Especially if he passes a disease on to the girls. Did you see the photos of those round worms? Disgusting."

In the course of all my phone calls and emails to MyPetChicken and postings at BackYardChickens, everyone had warned me to bring a towel with me when trying to approach my rooster, and to do so at night, when the flock is on the roost. Just toss the towel over that bugger, hold him down, grab his feet, and pick him up upside down. That'll calm him. Just don't hold him that way for too long, as he might suffocate.

I had never understood this violence. Big Red and I had an understanding. I'd Vaselined his comb on many an arctic cold night, hoping to prevent frostbite. He used to dance around me, his wing lowered, flirting. And I'd pet his back. Respond to his guttural sounds with coos and sensible discussion. I had never had a need for towels. In fact, other than Ping, he was the only chicken who would let me pick him up without any flapping and squawking. Thus, on October 13, 2015, I went out to the coop, picked up Big Red, and settled him in the hay-filled cat carrier.

As it turned out, he had a wound under his wing, thus, mystery solved, he could not fly up onto the roost. The vet inspected him, rather roughly, in my opinion, but confidently. He noted how thin the old boy was. He was concerned that Big Red didn't have enough body fat to fight off the cold and whatever else was ailing him. I noted that, indeed, he had lost a lot of weight. And that he was only three.

"Young," the vet nodded, frowning, feeling about for tumors or obstructions. "Probably worms." He suggested we bring Big Red into the house until he regained his strength.

Nearly making that suggestion moot, Big Red rattled and swooned. The vet's assistant and I looked at each other, wide-eyed. Big Red became a seemingly lifeless rag doll in the vet's hands. A shot of vitamins. Some de-wormer. We stood looking at Big Red lying on the table, eyes fluttering, too weak to move.

"Is he dying?" I asked. Big Red jolted, stood up.

"That's the vitamins kicking in," the vet said with satisfaction.

Carl has been known to ask me, "Who is in charge here?," thereby suggesting that once again I have gone over the line of chicken/cat versus human rights. Although I might have argued that his use of the pronoun "who" suggested that he, himself, had raised those other sentients to the level of being, for my own pride's sake, I was open to the suggestion that, of course, *we* were in charge. Only wasn't it our moral duty to be, at least, aware of those other creatures' needs, even if those needs were announced on an hourly basis through the night by a meowing cat who knew how to open doors?

It might have been our choice, but when I came home and announced that Big Red was moving in, Carl expressed his conviction that this was the beginning of a downward pitch on a steep slope. I bustled about, ignoring his negativity. Big Red moved into the guest room. In a dog cage filled with hay. Covered with a blanket. The cage, not the chicken.

After a good deal more googling and debate, I proceeded to deworm the girls, putting Wazine in their water for a day. I wondered if they had actually drunk any of it, but left it to fate. We threw out their eggs. Not that any of them were laying. They were too traumatized by Big Red's illness and subsequent absence from the coop.

Having him in the house was not the same as having him staring in via the front porch as he had the first year up at Darwin's View, crowing in at me, attentively studying me. Going to sleep that first night, with Big Red directly below us, it felt odd having him so near. And with hay showing up throughout the house, and the smell of chicken manure, it was not quite so comfortable and un-stressful as I had imagined it would be. My fantasy of how much calmer I would be if only the girls were inside, away from the cold, the heat, the bobcats and fishers? Never again. I was glad he was inside and warm. And I hoped he was getting better. Quickly. The girls missed him. He clearly missed them. Chicken dander accumulated.

Day one, we carried Big Red out to the porch so he could see the gals pecking about and they could hear his occasional, weak crow. I was writing less than I was searching online, trying to figure out what was wrong with him. Someone came up with the idea of gout … um. I moved on to read about worms. There are a lot of worms out there. Not earthworms, but disgusting worms. I had an awful suspicion that maybe it was a different

155

type that Big Red, and by association the girls, had, worms not killed by Wazine. We might need to use a goat de-wormer, Safeguard. Fenbendazole was much stronger and would take care of those pesky gape worms, hideous worms I hoped never to meet in person.

We put cayenne pepper and garlic powder on all the treats we gave the hens. Both promised to drive the worms out the back end instead of the front. It did nothing for the stress level, neither theirs nor mine. I wished Big Red would just get back to crowing and the flock to normal.

Day two, we took him outside, still in his dog cage, so he could get some sun and the girls could see him up close, if not personal. The girls *ran* across the driveway to him, clucking and churring. For the first time in too long, Big Red crowed a full cock-a-doodle-do. Not his lusty crow of yore, but a crow nonetheless. Little Big Man peered at him from across the drive. His feints at crows cowed as the girls crowded around Big Red. They didn't leave his side all day until, as the sun set, I took him back inside for the night, and they wandered back to the coop.

At four in the morning, Big Red woke us with a relatively lusty crow and I got the distinct sense that Carl had had enough. I took the initiative and suggested maybe Big Red should go outside again. He seemed to have regained his perk.

We carried the dog cage outside, opened the door. Big Red stepped forward and across the driveway, reentering his realm. The girls swarmed him, squatting in hopes he'd perform his duties. Normalcy at last!

Leaving Sarah and Peyton in charge of all things Darwin's View, Carl and I left for family commitments in New York City via Providence where we stopped by the urban farming store Cluck! I, as ever, struck up chicken conversation with the woman working there. I told her about Big Red's condition, the various prognoses he'd been given. And what would we do if we had to put him out of his misery? I mentioned taking him to the vet. Carl said we weren't going to pay a vet to put down a chicken.

"What else?" I snapped. "Lop off his head like we did his three sons? I do not think so. Not Big Red. No. Not him."

The woman intervened with the suggestion of a broom handle.

"Excuse me?" I said, staring at her, a sweet-faced, blonde-haired young woman who, apparently, had had a sick chicken herself. Her favorite chicken had gotten sick. She had to put her down.

"You hold the chicken by the legs," she said, "upside down to calm it, and then lay it on the ground. Place the broom handle gently on the neck, put your feet on either side of the neck, on the broom handle, and pull up vigorously on the legs. Instantaneous."

Apparently, there is a little bit of muscular reaction but no running around, no blood. I smiled and thanked her. We paid for the grease pencil that we would use to write on our metal vegetable tags so that the words LUFFA and EGGPLANT, BRUSSELS SPROUTS and KALE would not wash away in the rain—thus we would know what was planted where—and headed out to the car. I shut the door and looked over to Carl.

"No. Fucking. Way."

When we returned to Darwin's View, Sarah and Peyton greeted us. Peyton patted my hand. Sarah flushed and cleared her throat. She had not wanted to worry us. Me. Big Red wasn't doing so well. In fact, she was surprised he was still alive. She noted he hadn't been able to make it into the coop.

I put on my chicken boots, grabbed a flashlight, and headed out to the coop, where I counted. Eleven birds on the roost, in the nesting boxes. Indeed, one missing. I closed the door and went over to the southern run. Big Red lay sprawled on the ground, but alive. I knelt next to him, apologized for our too-lengthy absence, and picked him up as gently as I could, carrying him back to the coop and his hens. There I set him down, holding him steady until he gained his balance. As he had for weeks, he settled into the corner under the hens' roosts. Not, apparently, a protective act.

The next morning, he crowed. I had no idea how redemptive a cock's crow could be. How used to his morning medley I had become. If ever you have seen a rooster crow, you know the energy it takes. It's an all-body endeavor, the throat open and thrust forward. It takes heart. Soul. Rather like a musician. So it was no surprise to me that he crowed. Even if it were sung only in half. And then another attempt.

Sad plumage, what was left of it, and a bowed head. His comb was bloodied. The girls hovered, keeping him company. I picked up Big Red just as Carl came out with the car keys and cat carrier.

How to explain to a bunch of hens you are taking away their go-to guy? But, of course, they'd forget about him by the time we got back, right?

It is a gift to have a vet who appreciates a good rooster, and who will make time in his busy schedule to euthanize the same. I held Big Red while

157

we waited for the vet to be free, and imagined how different a life I would have had without him. I touched his head, his feathers. Remembered his transition from Rhoda to Big Red, his lowered wing and debonair dance for his hens, and me. His attacking Carl's drill, announcing treats to the girls, crowing to the sun, the moon, the world. Because of that crow, we had moved our new flock to Darwin's View and there they remained, and so did we. He was only three. This should have been his prime. I felt, at some level, deserted by he who had catalyzed so much, if only in my own mind. Death is so final. Breathtaking, and no going back. As with the death of his sons, I wished there were a way to go back.

When it feels like there is no choice, you just do it. Then you figure out what's next.

A shot of whatever it was in his heart. The vet said it didn't hurt. I know it did. Just for a moment. His eyes fluttered. He gasped. I held him as his body slowly let go of what was left of life. Whereas I had run away for the deaths of Pong, Cornelius, and Clayton, I held Big Red while he died. He was, after all, one of my original chicks. He had arrived a little ball of fluff. It was my responsibility to end his pain. I only wished I had realized it sooner.

Cancer threaded through his stomach and intestines.

When we got home, with Big Red in a box, the hens looked at me expectantly. They had seen me leave with him. Where was he? I couldn't show them his autopsied body. We buried him in the Chicken Graveyard with Chipper, Cheeks, and Opie.

With Big Red's death came the end of a fairy tale. I wondered if we had lived our good old days. Life wasn't happily ever after. It was more like that bear climbing the mountain, reaching the top, only to find there are more mountains to climb. Panda and Beatrice took on an air of suspicion whenever I approached. All the hens wandered aimlessly for weeks afterwards. Little Big Man attacked them, aggressive in his mating. Ping, terrorized by him, took to avoiding the entire flock. I would find her out in the rain, drenched. She lost weight. The others started to peck her. I thought she, too, was going to die. To give her time to recover, I introduced her to the dog cage, the porch. She liked it there. She settled in, quite nicely. Carl noted, after three days, that maybe it was time for her to go back out to the flock. But Little Big Man still chased her. I chased him. She had absolutely no interest in returning.

Extreme cold hit. We created a new coop situation, the Road Chicks' Quonset Hut & Bus Stop Greenhouse, and moved the distraught chickens in. For once, the human comfort of having only four feet to shovel and walk through the wind and cold took precedence over the chickens' comfort of staying in their familiar Chicken Paradise. The Golden Girls and Little Big Man's sisters, Peaches and Apricot, adjusted relatively quickly. Little Big Man, with his enormous and floppy comb, contended with frostbite and I with the guilt thereof. He and I had a bonding moment for me to put vitamin E on his dead comb. And Ping, the spoiled girl, my favorite, I reintroduced gradually, but still, whenever I went outside, she flew onto my shoulder to get away from Little Big Man. I missed the familiarity of Big Red and was convinced that Beatrice and Ping did, too.

Panda, CooLots, and Chickadee went about their business, pecking and dust-bathing. Beatrice remained bereft, standing in the middle of the run for hours, staring. She had filled out into a beautiful hen, no longer attended to by Big Red. She pecked a little bit, but a dark cloud hung about her. After all, I reasoned, she'd lost first her brothers and now her father.

"Ridiculous," people would say when I suggested she might be depressed.

One morning, Carl found Beatrice dead in the coop. The last of our Mutt Chicks and our last tie to Big Red.

Abruptly, the flock normalized. Ping adjusted to Little Big Man. No more moping about. The egg count jumped from four or five a day to six or seven.

The chickens, thus, left me behind in their adjustments to reality. Being human and thus more outside of nature than in, I was slower to adapt to this most recent episode of chicken drama at Darwin's View. Life just goes on after loss and death, which is part of its beauty, too, I suppose. Adaptation and survival. Because however much I think we won't survive the upcoming world that nature is gathering and forming, there will be some of us who might endure. I know this because, for years, I thought I'd sunk so long ago at boarding school. In fact, I'd swum, hard, lean, endless strokes of panic, loss, sadness, and rage. Starvation. Running. Tears. All formed a backbone of stubborn determination and survival. For better or worse, here I am.

IV

2016 AND BEYOND: DARWIN'S VIEW ONE BREATH AFTER MIDNIGHT

Dowsers point to the fact that Mount Monadnock lies on the divide of the Connecticut and Merrimack River watersheds. Water circuits and crosses beneath the rock and soil. Ley lines that connect ancient sites cross over and through this mountain. It is an energetically powerful place. A possibly healing place as the sun sets and the moon rises. The wails and yips of coyotes meld with the crashes of wind and lightning and summer's heat, creating a primordial scene: unseen shadows of sentient beings, black and blue clouds roiling, explosions of electricity webbing across the sky and connecting to the earth; trees bending under gale force winds, pelted by hail, sheets of rain.

If climatology studies are correct, which with every passing day and environmental disaster seems to be the case, then the nuclear clock, in the form of climate disruption, has ticked to midnight. The world as we know it is dying, and will be replaced by something other, something hotter, something we oxygen-dependent, temperate-temperature creatures cannot survive in. Had we only met our potential, recognized the gorgeous energy in ourselves and all around us, stewarded and cared for the vast web of this planet, how glorious it would have been—it was!—as seen through and experienced by *Homo s. sapiens*.

And Mother Nature? She will continue without us. The only difference between this sixth extinction and any other is that it is human caused.

LITTLE BIG MAN AND PANDA

Mud season is the fifth season in New Hampshire, occurring between winter and spring. It is a slurry shoulder season that harbors both winter and spring tendencies. That mud season, Ping and Little Big Man reached a truce and Ping returned to the flock to peck about merrily. Ever in never-a-dull-moment land, Panda began acting odd. As mother of the flock, she kept things running. She knew best. So when she started jumping on the other hens in a cocky way, I assumed it had to do with the pecking order and all things chicken. Maybe her hormones had confused themselves and this was a new way of going broody. I wondered if she would go broody. Instead, Little Big Man set his sights on her. He chased Panda around the yard, aggressive and relentless. All the girls looked ravished by his attentions, but Panda now provoked his ire. He would grab her neck with his beak, she squawking, lowering herself to the ground while he had his way. Brutal chicken rape. Even when he was done, he would peck and harass her.

This did not score him any points with me. The other hens, noting the potential, chased Panda, too, and one morning, the back of her neck was defeathered and bloodied. I set her up in her own private coop area—the Quonset hut with a large electric-fenced isolation area—and began to call around to rehome Little Big Man. No one and no bird would wreak havoc on my broody hen. The good news: mud season promised spring, and Friendly Farm, just down the road, was building up its flock for the summer. They were taking roosters, and Little Big Man was a rare breed. Carl and I tussled him into the black wire dog cage. He looked subdued and afraid as we drove him the five miles to his new home. I began to backpedal on my decision. Maybe we should keep him. He'd be scared in a new environment and no familiar hens to keep him company. He would miss his sisters and they would miss him. What if he became soup? Was this the right thing to do?

Then I remembered Panda's bloodied neck and his aggression.

Peyton, who worked at Friendly Farm that summer, reported that she would see him pecking about and, on occasion, getting into cock fights.

In response to my concern for his safety, she noted, matter-of-factly, that's what cocks do. She said he held his own and had his own favorite hens.

Our hens recovered from Little Big Man's attentions. I noted the silence of no cock-a-doodle-doos. Big Red's daily song, and Little Big Man's rendition of it, had heartened me. Hearing those cock-a-doo challenges heralded that the hens were likely alive and safe, and made my own endeavors against life's inevitabilities seem not so quixotic. However, I didn't miss the squawking of a cock's sudden attack on a hen. That silence left room for a dawning recognition. With no roosters to harbor, we could bring the girls with us for my mother's July visit to Rhode Island. The Road Chicks concept was (re)born.

That June, in between recording sessions in Rhode Island and Connecticut, Carl transformed the old Quonset hut coop into a road vehicle. He cut off the tongue of our trailer bed—bent so long ago during the relocation of the Chicken Palace—and moved the Quonset hut coop onto the bed. He proceeded to drill and hammer and connect the coop to the trailer such that it was a single lean-and-mean unit. He attached lights to the coop and

Panda

the trailer to the Subaru and headed out to drop the Road Chick chicken coop in Providence en route to his recording session.

That was the plan. The hens joined me as I watched the coop bounce and jump down the driveway. It had an alarming sway. Carl made it a mile down the road before the dramatic swaying convinced him to turn around. He returned to disconnect the trailer, then raced off, late for the recording.

While he worked at his recording sessions, I visited with my mother and sister in Connecticut. Upon our return to Darwin's View, our chicken sitter left and I went out to check on the girls. Panda was in the nesting box looking thoughtful. It being June, she was early to her usual late-July calling, but I returned merrily to the house to inform Carl we just might have a broody hen and wasn't that exciting? I debated what breeds to order even as I scolded myself because I was never going to order chicks again. Local chicks only!

I went out a couple of hours later to check on our mother hen. Her head rested at an odd angle to her neck, which lay flat and long on the hay. Hens don't hold themselves that way. She panted. I knelt to pick her up, hugged her as I carried her to the porch. The sun set as I sat with Panda in my lap, holding her closer and longer than I ever had. She was very hot, she who enabled us to have chicks without the worry of heat lamps. She who answered the chicks' alarmed chirps with a motherly crawking sound, hovering and educating them how to scratch, peck, be. I wondered if she were egg-bound. After all, she had been in the nest. Hens die of egg-boundedness. I would never forgive myself if she died and I could have saved her, and so I retrieved plastic gloves and petroleum jelly. With Google to assist me, I delicately checked to be sure all was in order. I watched Panda closely, as if she might guide my exploration. Her big, black eyes stared ahead of her. She exuded heat and breathed heavily, beak open. I had no idea what I was looking for but felt no egg. What else could I do? I settled her into the dog cage, having prepared it for another convalescence, and went to bed.

The next morning, I found that Panda had followed in Big Red's claw-steps.

Her loss was even worse than his. Her death made the loss of the other two—Mother Nature and my mother—conceivable. If Panda could die, maybe their demise was inevitable, too. It was one of those haunting events that I kept returning to, wanting to rewind time, to prevent. In all the stories about the chickens that I told myself, Big Red, on occasion, had heroically died, saving the others. But Panda lived. From her chick-dom, I had known she was the mother. Wise and queenly, bright-eyed, her black fluff becoming green-hued black feathers. Our black beauty, broody hen, the mother figure to all our chickens, played a life-giving, everlasting role in my heart and imagination. She was a key component to the future. She couldn't die. A lack of crowing and chutzpah was one thing, but no mother hen?

We put her corpse in the refrigerator for the weekend so that I could take her to the vet. Adding insult to injury, the autopsy showed she was a healthy hen, lots of good meat on her bones. Nothing was wrong with her at all. Except that punctured air sac in her lung. She must have hit or been hit. Chickens are quite delicate creatures. The slightest thing can trigger death.

So. Eight hens. Only two of our original chicks. We prepared to move back to Rhode Island for a month, during which time my mother would visit. What I had been wishing for—being in Rhode Island with my mother, the hens, the cats, and Carl—would happen.

THE ROAD CHICKS

As Carl has been known to say, any job worth doing once is worth do-ing twice. He succeeded in getting the Quonset hut coop road-wor-thy. Leaving the chickens in the care of their usual chicken sitter, we drove the coop to Rhode Island with the two cats. The next day, while the cats reacquainted themselves with their favorite napping places, I converted my office into my mother's summer bedroom and moved my desk and com-puter into the entryway. Carl studied the backyard and the coop situation.

"Why don't the girls just use the old coop?" he asked.

"Not enough room," I replied. Five young pullets and a cockerel take up way less space than eight full-grown hens. As we discussed the situa-tion more, we agreed that putting the coop in the garage made the most sense, if only because getting the Quonset hut into the backyard would be putting a camel through the eye of a needle, given the two-hundred-pound Quonset hut coop was five by eight and the fence door was three feet wide. Besides, the garage seemed safe. It was cement. I had no inten-tion of bringing the girls to Rhode Island only to have them taken out by one of the raccoons born the same summer as the chicks' adoption so long ago. By now, I was not blind to nature's sick sense of humor.

We leveled the coop, using cement blocks under the trailer hitch lip, then placed one end of a ten-by-one-foot board on a wood sawhorse level with the coop door and the other end on the window sill that opened out to the fenced-in backyard. Voila! A gangplank from the coop to the win-dow. I spread hay on the board for hen comfort while Carl fabricated a chicken wire cover so that they had a protected run up the gauntlet to the window that opened onto the yard where the original six had spent their youth. Their tiny old coop, with its Fort Knox-style fencing, still had a working door. I cleaned it out and filled it with hay, too, in case the acces-sibility of those nesting boxes tempted one of the girls who, in her passion, didn't feel like making the journey back up the steep ramp, into the garage, along the plank, and back into the Quonset coop to the nesting boxes.

Admittedly, it was a trek. But it was raccoon-proof. And it was what it was because my mother arrived and we had to settle her into the house, meeting aides and physical therapists and preparing meals. Mom. Carl. Cats. The familiar comfort of city life and Beachmont. Home. All that was missing were the chickens.

Two days into my mother's visit, Carl and I drove up to Darwin's View to get the hens. We busied ourselves tidying the yard and the house until dark, when we would remove the girls from their roosts, put them in the back of the car, and get home in time for dinner.

Another case of bad calculation. It was summer. We had just passed the longest day of the year. At five o'clock, our intended departure time, the sun had no intention of setting; nor did the girls show any signs of getting ready for bed, unlike me who would have tucked in quite happily by seven, which clearly would not be happening because at five-thirty, we still had a two-hour drive ahead of us, not to mention the introduction of the hens to the newly refurbished Quonset hut. After a brief discussion—who is in charge here?—we decided not to wait until the hens were ready. We would corral them into the car. Why had we brought the Prius and not the Subaru?

We put down the rear seat in order to have full use of the cargo space, which we proceeded to fill up. Two ten-gallon buckets, one of feed, the other of cracked corn and mealy worms. A fifty-foot skein of electric fenc-

ing, folded up into a barrier between the front seats and the back. A chicken waterer. A tarp to line the back of the car so as to avoid any long-term effects of chickens and their manure. A bale of hay, spread out for comfort in the now rather small area left for the eight hens. Sarah arrived—happily for us, maybe not so much for her—just in time to help us catch the chickens, who were exhibiting their usual feral traits.

With the help of mealy worms, we got the girls into the northern run. Eight hens, Carl, and me bumping into each other in the run, and

Road Chicks: Brownie, Clownie, Downie, and Ping

Sarah working the door. We got the easy ones out of the way first. The perky Ping, then Chickadee, plump and sweet and quite surprised she was included in the game of catch-me-if-you-can. CooLots, being the large, matronly hen that she is, and perhaps assuming herself exempt from the insult, moved slowly. The blondes were determined to foment a riot, and the babies, Peaches and Apricot, exhibited their phobia to touch. I wondered what had instilled in them such a terror of us. All we did was arrive with food and water and free them on a daily basis, and in return? They refused to let us come anywhere near them. They ran. Feathers flew. Somehow, with no detached wings, I'd catch a hen, hand her over to Carl, who would hand her to Sarah. One or both of them would carry the protesting prisoner to the car. The first to be caught was slipped in through the back hatch, and the last—after those already inside began pushing and shoving to be free—through the window, up and over the electric fencing.

It took nearly half an hour to get them all in. As I released the last one into the back of the car, one of the triplets, attacked by her sisters, flew into the electric netting. I reached in, untangled her, counted eight hens, and plunked myself in the front seat.

"Let's go, gotta go!" I said, holding tight to the distraught Clownie, who was apparently rethinking the pros and cons of being pecked by her sisters in the back of the car now that she was in the hands of a human. Too late! Covered with my usual accessories of chicken feathers and dander, I slammed shut the passenger seat door. The sun hovered over Mount Monadnock. Sarah waved us off.

As we drove down our half-mile-long drive, I looked back and was struck by the abandoned quality of Darwin's View. The wind blew dust and dirt about and the scene felt very much like it was plucked from a movie about a mysteriously deserted town. The vacant house. The barren driveway. The empty coop. No pecking and clucking. No heartbeats to liven it. I missed the chickens even as they scratched in the back of the car, and I noted that, for the first time in three-and-a-half years, Darwin's View would be uninhabited. It felt cold on that summer eve. Abandoned. I wondered what the future held and, as we stopped at the mailbox to pick up our junk mail, had a bad feeling. Junk mail. I shoved a couple of envelopes under Clownie just in time for my premonition to come true: a splat and squirt, and the stench of chicken manure. Carl commended me for my forethought. We stopped at a gas station in Jaffrey to get rid of the soiled envelopes and proceeded on our way.

The girls settled down, though not Clownie, who panted from stress the entire trip, alert to my backseat driving even after the sun set. We drove in darkness but for the highway lights and passing headlights of other cars. She stared out the window. So did I as the landscape transitioned from country to urban, more and more cars, too many cars, and what kind of brain-trust idea was this to drive pet chickens over state lines? As ever happened when we reached the Rhode Island border, something flipped in me. My thoughts shifted from the To Dos in New Hampshire to those in Rhode Island. Darwin's View left a stark, cold impression on my soul, whereas the Beachmont house felt cozy. That's where I wanted to be. Maybe all the transitions had worn me down. I needed one place and that place required only Carl, my mother, the cats, and chickens.

After we arrived, Clownie was the first into the Quonset hut, the first to settle in. The others, sleeping now, were easily moved. Thus returned Ping and CooLots to their home on June 25, 2016, four years to the day after they had first arrived there as peeps.

The Road Chicks might have earned their moniker, but no pirate games for them. The next morning, they refused to walk the plank. I carried our adventurer Ping to the other end of the gangplank. She looked about and hopped out into the fresh air and sunshine without a backward glance. We encouraged the others to follow her, but they remained huddled in the coop. I suggested to Ping that she come back and tell them all about the big pile of leaves full of worms, the shaded areas with hostas to dig up, and the sunny areas with clover and dandelions. Ping was too busy and left it to the humans to plead with and eventually carry her sisters, one by one, to the end of the plank. No. Once out of our embrace, they would bolt back into the coop, ever more leery of getting caught again, and we nervous that they'd get out of the coop and chicken-wired plank and into the garage with all its filth and clutter.

"What are we going to do?" Carl asked, frustrated by yet another morning going to the dither of chickens. I smiled. It was time for our very own nuclear option.

"Mealy worms."

Did the girls perk up at the words, the way a dog does when one spells out W-A-L-K? Desiccated fly larva. That's what the girls wanted. They gobbled and clucked as I strewed mealy worms along the plank. They pecked and walked, then froze. More mealy worms, and not a few pushes and shoves through the wire, and shutting the coop door so they couldn't get back in, and eventually they were freed to the great outdoors.

Our breakfast earned, Carl and I returned to hang out with Mom and Denise. I made espresso drinks and worried about the raccoons. I went outside to verify that there were still eight chickens. Seven. I counted again. Seven. I called for Carl and raced out to the driveway. Sure enough, there was Apricot. She dodged me. I chased her, vowing not to let the chickens ever range free again. I would spend any and all of my free time with them. They would learn to trust me. If only Apricot would smarten up and get back to safety. The alternative analysis of the situation was, of course, that I would smarten up and leave Apricot free to live her life, however foreshortened it might be.

Carl arrived to help. Together, we managed to chase that little hen across and down the driveway, around the entire exterior footprint of the house, and back into the fenced-in area, where she joined her sisters. Carl returned inside, not a little disgusted with how the chickens ruled our roost. I recounted. Seven. This time Peaches had escaped. She had flown into the neighbor's yard, using the platform outside the garage's gangplank as catapult, and made it clear that she had no intention of being told what to do or where to go. She would fully explore her environs, or die.

That seemed a likely scenario when she ran toward the road just as a car approached. I went wide and came at her from the street. She veered to the gardens, dodged me, and ran across our driveway to the hedge that stood between our property and our neighbors. I really, really hoped they weren't watching me chase a chicken around their yard, then under the hedge. I knelt down and reached for Peaches, who stood just out of reach. On my belly, I crawled under the hedge. She casually walked back onto our driveway. I played defense, offense, and whatever else—and no sign of Carl. Did he not notice my absence? Wasn't he curious where I might be? Under the hedge again, then out of it, the hen and I danced a *pas de deux*. I suspected she looked at it as a game. Finally, Denise and Carl—who

170

had, apparently, enjoyed a pleasant coffee and read the news while I played hide and seek with Peaches—came out. One hen to three humans. I felt heartened by the ratio. Indeed, together, we successfully herded the hen through the screened-in hallway between the house and the garage, and out into the chicken garden.

Carl and I took down Freedom Platform, and spent the next hours devising a new approach up to the garage window and the Quonset hut's gangplank. It looked great, if narrower than CooLots and Chickadee.

At dusk, I went out. No chickens in the yard. Creatures of habit and ever aware of their circadian rhythms, they had put themselves to bed. I headed into the garage to wish them a good night, congratulating myself that they were in such a secure situation. The Quonset hut coop inside a cement garage: Hah! I dropped a gauntlet to any raccoon who might dare attempt the ravagement of our chickens' Alcatraz, and peeked into the coop. Empty. Not a hen to be seen. I verified that their plank to the coop was still in order, that they weren't huddled anywhere in the garage. Alarmed, I went back out to the garden. Cooing and rustling emitted from the old coop. I unlocked the door. Sixteen beady eyes and a wall of feathers, stuffed in and panting because it was a hot night in the city.

I wasn't going to argue the point. Ping and CooLots were the ring leaders now that Panda was gone. They had only followed the old rules and habits. I knew how com-forting the familiar could be. And this way they could come down into the screened-in run area every morning to stretch and look at the day, instead of having to wait for us to let them free from the comfortable, spacious, safe situation in the Quonset hut that Carl had spent days adapting for the road and that we had spent hours setting up, ever with their

The girls on a heap of compost in Providence

The start of the addition

best interests in mind. Yes, this was much the better choice, to be all stuffed into a coop made for six. I made sure the doors of the coop and run were secure. An egg in the nesting box the next morning proved the girls had settled in. They pecked and scratched in the backyard just as had our original six. They stood on the raised herb bed, preened, and stared in at us, sitting in the screened-in porch, their very own arcade video game.

I went outside to join them, glanced at Ping who studied me, her rose-combed head ever-so-slightly tilted, expecting correctly that I would give her a treat. I held out a grub to her, she a heartbeat of continuity and acceptance throughout all our changes. Change happens. Our creative force can take a blob of clay, as my mother used to, and form it into one's vision. At times, it fails to match the imagination. But when it does? It's like a caterpillar dissolving into a soup of enzymes, later to emerge as a butterfly, or a tadpole's transformation into a frog.

Time passing. Adapting. Nesting. Carl and I will talk about all sorts of ideas. We hash them over, debate, tug and pull, poke, and then, one day, we push the "let's do it" button. From the outside, it looks like spontaneous combustion. In fact, it's a natural sequence of events. Like a move to New Hampshire for a winter. Another renovation of a chicken coop. Or putting an addition on our off-grid house, installing a natural swimming pond, and commencing permaculture site work, all at the same time. And when those fifteen months of living on a construction site were done? We set our sights on Rhode Island.

Our Beachmont home had become a way station for others to nest in until they were ready to fledge. Katie, our house, cat, and chicken sitter over the years, still stayed at the house one or two nights a week, using my office as a bedroom. My cousin Lise and her husband, Miles, had lived there for a year while they transitioned from Michigan to Vermont. In their place, Paige, the daughter of an age-old friend of mine, and her husband, Michael, had moved in. They hoped to save money to buy a house of their own someday. It was a gift economy. Everyone gained from it, especially Carl and me, who could sleep easy at night knowing the house was tended to, and still, we could drop in for an overnight or a pause before I caught the train to New York City or upon my return. And then there were our day trips.

April 4, 2017. My mother was scheduled to arrive in Jaffrey in a couple of days to visit for a few weeks. Carl had an 11:00 a.m. dentist appointment in Rhode Island. We headed down. My cell phone rang as we approached Worcester.

"Tory?" Katie said. "I'm at the house. I'm really sorry. I just took a bath in your bathroom and when I was draining the tub, the toilet started to gurgle. And then shit started coming up in the shower."

The joys of home ownership. I looked at Carl. He looked at me.

"We're on our way down anyway, Katie," I told her. "I'll call Roto-Rooter."

While on hold waiting for someone to help us, Carl sang along with the Roto-Rooter theme song. I found it fascinating how low his voice could go.

"It's kind of a no-brainer," I said.

"What is?" Carl asked, just as the Roto-Rooter man answered. I explained the situation. He would meet us at the house within the hour. Once off the phone, Carl and I were silent for a moment.

"It's a lot of money to maintain it," one or the other of us began. It didn't matter who. We were both thinking the same thing. As we talked, assessed, anticipated the cost of our meeting with Roto-Rooter, we crossed over the border into Rhode Island. My heart remained in New Hampshire.

Mr. Roto-Rooter arrived minutes after we did. He scoped the pipes and reported that there were roots growing into them, blocking the flow. He would have to do a full scope. The cost would depend on the blockages and where the main was sited. In the meantime, we shouldn't use the water until the pipes could be fixed. Roto-Rooter was busy, given all the rain; what had been a blizzard in New Hampshire had been rain in Rhode Island. They wouldn't get to our pipes until the weekend. It was Tuesday. He chided us: his boss had been to our house a couple of years earlier and told us to fix the pipes.

"I don't think so," I countered.

"My boss says—"

"Call your boss," I snapped, feeling the blood rise to my cheeks. Maybe it was the dawning of what had happened and what would. Maybe I sensed what the next months held. What was certain was this: when it comes to homes and renovations and the workings of a house, Carl and I might go overboard but we do things right. "Get the date. If we were told to fix sewer pipes, we would have fixed them."

He called his boss who looked it up. The answer: three days before we had bought the house in 2010.

"Right," I said. Pyrrhic victory might come to mind. I asked the man to please schedule the house's equivalent of a colonoscopy.

Sometimes there's no going back.

Carl left for the dentist. I called real estate agents to interview.

Within five minutes of her arrival at Darwin's View, Mom and Carl were sitting at the piano, she leaning against him while he plunked out a tune. They warbled the words to "My Funny Valentine," "Somewhere Over the Rainbow," "When You're Smiling." Denise unpacked. I got dinner ready, savoring the perfection of the moment: my mother's presence in my home. Communication might be imperfect with words disappearing and sentences trailing off without end, but, at least, I could help her walk to the couch. I could watch her sleep, or, together, we could watch a movie. I could feed her, kiss her, hold her.

Mom with a peep

A week or so into that visit, we all tumbled into the car to go on errands. Pick up happy milk from our biodynamic farmer friend. Olive oil and vinegar at the shop that happened to be next door to a local ice cream shop that had Totally Turtle caramel ice cream (Mom's favorite) and Oreo Cookie . . . no, Mississippi Mud . . . no, Kahlua Brownie (mine). We proceeded to Agway for our happy chicken feed. There we found hints of spring as Carl and I entered and headed to the feed area. Mom and Denise stayed in the car as it was a quick stop. Carl and I walked by the muck boots and soil amendments. Tulip and daffodil bulbs. Seed packets. Tubs full of cheeping chicks. They were really cute. Eight or nine Speckled Sussex chicks skeetering about, one falling asleep with her head in the chick crumble. Stunning but true, Carl passed by them without a hello. I stopped to stare and cluck, taking note of the Do Not Touch sign. Carl came back with the feed bag slung over his shoulder to find me still there. I looked at him. He looked at me. We went out to the truck to announce the happy news to Mom and Denise.

By the end of the day, and a trip to Keene to get more of a *variety* of chicks, we had adopted six adorable *girls*: two Speckled Sussex dubbed Squeaky and Splotches; two Golden-Laced Cochins, also known as the Suffragettes Susie B and Cady due to their feathery, pantalooned legs; and two Buff Polish. These last two, who looked particularly silly with their cotton-ball topped heads, we named Muffy for her muffin top and Moey for her Mohawk. Peep, peep, peep. The chicks scurried about their plastic bin next to the woodstove and provided entertainment for us, if not for the cats, who would watch them for a short while, then yawn and go off for a nap.

I hadn't played the part of Mother Hen for five years. I watched them flap their tiny wings that sprouted feathers to cover the fluff of their infancy. I admonished those who were too forceful in their practice of the pecking order. I watched them grow up. They grew so fast. I enlarged their quarters by putting them into a box. Not long after that, I noted to Carl that it might be time to move them into the greenhouse. Just a corner of it. One that could expand as they did. He didn't respond, more focused on an upcoming gig and the Beachmont house plumbing and heating systems, both of which had to be fixed by early May when the real estate agent wanted to put the house on the market.

I took matters into my own hands. While Carl was at his gig, I put newspaper on the stone floor of the southwest corner of the greenhouse. This would protect the stone from chick goo, and perhaps insulate the chicks' claws from that cold stone. I put cardboard along the windows of the greenhouse to protect the littles from too much cold at night, too much heat during the day, and possibly killing drafts. A thick covering of pine shavings on the floor. An old (okay, new, but we weren't using it at the time) screen door as a barrier between the chicks and the cats. One of No-

Nick and Nora admiring the chicks

ra's discarded cardboard boxes placed in the midst allowed for shade from the heat lamp that I attached to a plant stand. An old tree branch served for their practice roosting pole. Grit, food, water. And then the chicks, one to six. They set about exploring their new and *enormous* surroundings. Loud chirrups and squeaks. A lot of practice flying and perching on top and inside the box. Ten minutes later, they were sprawled, exhausted and asleep in a pile. Carl, for his part, was surprised, upon his return, to find the greenhouse lit up by the heating lamp and the chicks ensconced therein.

They did grow. Soon enough, they were old enough to have play time, just for an hour or two, outside in the old portable chicken run that used to be the Hay Chalet, and I had moved their inside space to the southeast side of the greenhouse, into the four-by-four-foot bay next to the door out. Nora would nap in a sun spot, her furry cheek leaned against the screen. All she needed to do to oversee all chick activity was open her eyes. On occasion, she would stretch her claws, yawn again, open her eyes wide. Another week. They expanded to another bay, then another, by which time we were ready to move them into the Quonset hut coop, which happened on their two-month birthday, a celebration of their gawky adolescence.

The night before she left, I went into my mother's room to say goodnight. The nighttime aide was bent over her, trying to understand what she was saying. My mother's face shone with the glow of a special secret. She held something in her hand. The aide made way for me, and I sat down next to my mother.

"Here," she whispered. "Do you see?"

She opened her precious hands, curled like paws, and held them up to me.

"This was Nana's," she whispered, holding what she saw with care and reverence. She tremblingly folded it into a Kleenex. My mother adored her Nana, her mother's mother, who went to college, collected Appalachian music, and survived the loss of her young doctor husband. Strong, smart, loving. A bulwark in my mother's childhood.

"Do you want me to take care of it?" I asked, whispering, too.

"Yes," she whispered. "Yes."

She held the wrapped tissue up to me. I took it from her just as carefully, reverently, as if it were a holy object. It was. I stood up and looked around, debating, then turned my back so she couldn't see where I put it. I opened the Kleenex and, because where else could I put such precious cargo, I held it against my heart. I hold it safe. It will be there if she ever needs it, and when I do. Because someday that which I have dreaded all my life will come true.

This is how I learned a new language.

My mother left, back to New York City. Carl and I began the Beachmont house staging process. We drove back and forth to Rhode Island, packing and boxing, tidying and throwing out. What to take, what to give away, what to stage?

I skeetered over the emotions of loss and letting go. Habituated since childhood, I ever held the mind above the heart. The heart knows best, but only if you let it. I tussled with pain, but didn't sit with it and listen. I left my heart to process in secret the deeper consequences of what we were doing, while my mind logically determined my next steps. Both heart and mind skirted around the question *Why? What was I doing? Should* opposing *want to.* I told myself pretty stories like *being.* Like *not needing words to be with someone.* Yes, pretty stories of having a bigger purpose diluted the love of place. And ever deeper, a more distant call echoing, *Can't I go back?*

Carl and I put our hearts' sweat into our homes, something real estate agents don't care about. And we know that there are houses for sale, and people who want to sell houses. We have never asked a real estate agent to make us lots of money. We always ask them to price it right so that we can rip off the Band-Aid, be done with it.

"It's really hard to price," all the realtors we interviewed said. "It's a spectacular place. You've really done amazing things to it. You should get top dollar."

"Please don't overprice it," we replied, limp with physical and mental exhaustion. "We want to sell it."

And so the professional priced it, held many open houses and numerous walkthroughs. Everyone loved it. No offers. We lowered the price. Twice. The third time, finally, a flurry of offers. All but one fell through. The closing would be September 6, 2017.

Taking a break from unpacking in Jaffrey for some adult chicken time, I noted that Brownie had lost her under-feathers. Her stomach was completely naked and bright red. Other than that, she seemed perfectly fine, racing about, pecking within the order of the pecking order. Then, one morning Carl came into my office to say that Downie was dead. No, not Brownie. Downie.

178

We had no idea why, but buried her immediately, rather than risk finding her body forgotten in and amongst the avalanche of stuff beginning to clutter the garage, the basement, the house. At which point, I noted that Clownie seemed not so well. She would let me pet her, instead of skeetering away. Later in the day, she would be fine and happily dodge my attempts at touching her. I chalked up her indisposition to indigestion. But what about Peaches? Though still laying her blue eggs, she, too, seemed under the weather. She huddled in a corner, obviously uncomfortable but racing away if I approached. I couldn't get near her.

One night, early in July, I was putting the girls to bed. It was after sunset and so they were relaxed and dozing. I noted Peaches on her roost, studied her more closely. To her surprise, I picked her up, shone the flashlight of my camera on her: she had a huge, crusty growth on her neck and another on her tail bone. What kind of animal caretaker was I if I allowed my girl to get covered with I-knew-not-what and nothing online told me more? The next morning, I put both Peaches and the featherless Brownie into cat containers and drove to our vet.

Peaches' cancerous growths blended nicely with her feathers and down, the vet noted. He assured me it was easy to miss. I didn't believe him, but agreed that it made no sense to keep her in that discomfort known as pain, which she was in. He took a scrap-

ing from Brownie's underside to send in for testing while I said good-bye to Peaches. And then he did the deed. I took Peaches home in a box. Two days later, I returned with Clownie, who had taken a dire turn, tail down, not eating.

Cancer, too, and he was convinced it was a result of a Vitamin A deficiency, especially when Brownie's skin test came back with exactly that result. I questioned him: We fed them organic feed that had a daily quota of Vi-

Peaches

tamin A. They were free range. They got daily treats of kale and sprouts. How could they be deficient? The vet shrugged.

Carl and I buried Peaches and Clownie with Downie and the others in the chicken graveyard. I studied and worried about Brownie, who continued to gad about, apparently indifferent to whatever skin thing she had going on. I grumbled about the healing qualities of the place we called Darwin's View and contemplated genetics. Of all the chickens we had stewarded, twenty-six in total, we had lost

The Stoner Dudes: Mo and Muff

five to cancer. I looked at the still-living eleven and wondered . . .

At which point Mo crowed. One of those sad, out-of-tune cockerel crows. I hardly noticed it, it was such a natural sound. And then I did. I ran to tell Carl that we had ourselves a roo again, focusing on the positive rather than the negative of a cock-a-doodle-doo.

One roo? Two. Mo and Muff, who loved nothing more than to sit side by side on a fence, their crests of feathers hanging over their eyes. They looked like of a couple of dudes just hangin' out, playing it cool, all peace, love, and harmony. We named them the Stoner Dudes. Gentle, slightly neurotic, and calm—until they weren't. They began to practice laying. These dudes were not what could be called agile. It was embarrassing to see them jump on one girl or another, only to fall off, flustered and stamping in circles. Mo developed faster than Muff and began to cross over to the older hens, easing his way into their circle, tossing his crest feathers back, like a Fonz wanna-be, so he might see. The two flocks free-ranged, but the pullets and cockerels gave wide berth to CooLots, Ping, Chickadee, and Brownie. Apricot, always a shy and retiring girl, was about the pullets' size, and although she occasionally pecked to maintain her position on the pecking-order ladder, she tended to hang out with them as much as with anyone.

But still the two flocks were two, not one. It was early August. Although busy packing up the Beachmont house for sale, and ferreting through all the clutter that had filled the addition to the brim, I still made time to worry and wonder how I would ever get all those feathery personalities to come together for winter. Every night, the old hens would go into the big coop and the pullets into their Quonset hut, and, although on one occasion Squeaky ended up in the big coop, squeezed behind a hay bale for protection, there didn't seem to be a coming together anywhere in sight.

Tick, tick. Mo got cockier and more confident and took less guff from the older girls, who looked at me and then at the boys as if to say, *Really?*

Maybe for the older hens, as for me, none of the roos came close to the kingly memory of Big Red. He had been the ultimate roo with his proud swagger, glorious orange, red, and brown plumage, and lusty crow. Yes, he'd had his moments of being less-than-brave, like that time CooLots ran from the hawk and Big Red beat her into the coop. And he'd had humiliations, as when he had joined Brownie, Clownie, and Downie down in the compost bins. They'd all pecked and scratched about, but the girls got bored and left. They were petite and flyable. Big Red was all bulk. He clucked and scolded alone in the bin until I realized he couldn't get out.

The two of us pretended otherwise, of course. He had to keep his pride in front of the girls, who played their parts, too, pretending ignorance. While Big Red authorized me to pick him up out of the compost pile, the girls looked to the west, facing the breeze. With utmost respect, I carried him to a high place and set him on a hay bale so that he might strut and crow his disdain now he was back to freedom. The girls came running, clucking at him as if questioning, *Where have you been?* And he, as ever, chortled, announcing that he had found a treat. Here, take it. Because always and ever he took the hens into account, sharing treats and celebrating their eggs.

That being then and this now. There weren't many eggs at this point, just one or two a day. Our old hens had, apparently, had their heyday of egg-laying. Things began to get a lot less peaceable. Initially, Muff hadn't minded playing the subservient role to Mo's *I am the King, I rule because I have the best and most floppy feathers falling into my eyes* role. Muff seemed quite content, grateful just to be included. But time ticked and my hopes of keeping them both dwindled. Mo got more aggressive and then Muff did, too. They began to tussle. I tried to take a positive attitude. That was my new mantra. ACT! Attitude Change Time. Be positive. Good things will come. And they did.

One night, I went to put the big girls to bed, and lo and behold there was Mo on the roost, huddled in the corner, looking more than a little tense but determined to stay. He had figured out the laying thing, and apparently had decided that was the night he would become head cock-a-doodle-do. I was so proud of him! What bravery, to sacrifice a night without his little sisters huddled about him, shoving to be nearest, opting instead for the cold shoulder and an occasional irritable peck.

The next day, I noted that Mo was bringing the two flocks together. His sisters were nervous but willing to scratch and peck near Apricot. Ping shared a bit of gossip with the Suffragettes. Squeaky and Splotches hovered around CooLots, teasing her because they knew they could outrun that grumpy old hen who yet found the best treats. Muff kept to the background, hiding behind a piece of plywood that leaned against the garage. We could hear him practicing his crows there, something he never did when Mo was nearby. I suspected the resounding echo reassured him, and knew we had to find a home for one of those roos. Our main coop might hold what was left of five generations of our chicken trials, but it couldn't hold two roos. A too-familiar dread and anxiety filled me.

After mentioning to the Landises that we had an extra roo on our hands should they happen to have a shortage, they looked thoughtful. Later that day, Terry called to say that, in fact, they had just lost their wonderful, gentle roo and were open to adopting one of ours. Happy days! We have the best neighbors in the world. Happy night that they came to pick up Mo. He was the most beautiful, and I thought it appropriate that he rule over more hens than less. He settled right down into Terry's lap in the truck and went back to sleep for the ride over the hill. The next morning, I could hear him crowing, and Terry texted to say he had settled right in and their hens liked him.

I contemplated Muff. Now Mo was gone, he tested his vocal cords out in the open, pretending bravery in a panicked way. He followed his sisters about, taking their lead if they ran, because he couldn't see above him. I had to wonder about breeders who would breed a chicken that couldn't protect him or herself due to an overabundance of feathers clouding the eyes. Muff developed a last-minute dodge technique, seemingly oblivious that I was anywhere near, until there I was! Run! I renamed Muff "Schtude," an affectionate name for a goofy roo who startled every time I came near and, someday, I knew, would be Uncle Schtude.

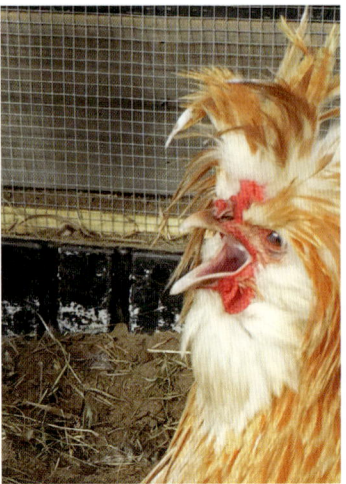

Schtude

AND NOW?

There is something deeply stressful in the cacophony of a hen's announcement that she has had an egg, and then Big Red's . . . Cornelius's, Pong's, Clayton's . . . Little Big Man's . . . Mo's and Schtude's call and response. It is a vital life sound. An urgent, gut-pulling, uncomfortable sound. Rather than celebration, it seems like an announcement of danger and disaster. A rooster's crow has a desperate edge to it, as if he's trying to say, *do something!*

My mother said to me, "You have the time. Do something."

This being counter to what Buddha might have said: "Don't just do something, stand there."

Monadnock means a mountain that stands alone, isolated. It is also the mountain that stands before me every morning when I rise and look out the window to the west. I suggest to Carl that we climb it more often. By now, he knows my ideas are bigger than my abilities. But in my dreams, I climb it every day.

My life has been a seeking for a lake with blue flowers, each day underlined with four-letter words expressing absence, not safety: loss and fear. Want. Risk. A lifetime of guilt and shame at having so much when others have so little, of denial of myself because I did not deserve. But I am at the start of a new path, a journey deeper into the heart, to the beingness that I have avoided. I am not there yet. But at Darwin's View, under the sweet glow and darkness of the moon, I am developing a new list of four-letter words: Soil. Heal. Grow. Love. Home. Here, I have found my lake. I call it a pond. It is a pool I swim in, west to east, between Mount Monadnock and the Wapack Range. The water is cold to the point of breathtaking. The resi-

The natural water swimming pool

185

dent frogs cheer me on as each stroke makes me stronger. When I come up for breath, I note the plants of the pool's stone bed regeneration system. Blue Flag Iris. Bearded Sedge. Monkey Flower. Pickerelweed. Below me, water bugs do the breast stroke. Snails on the side of the pool. A dragonfly birthed from its cocoon. Skin chilling, I feel the shape of my body, where it ends and the water begins. My awareness of my feet, legs, torso is vivid. The water livens me. It heals me. It is me. Being. I am in the moment because the moment, this one in the pool, is absolutely perfect, and being here at Darwin's View is exactly where I am meant to be. After these last seven years of bewilderment, the word gratitude takes form. Happiness. I will not look back. Instead, I will plant blue flowers. Delphinium. Geranium. Forget-me-nots. And I will go step by step, not looking too far ahead, and being sure to hydrate. And then I will look about me and breathe. I no longer ask, where is home. I ask, how might I save it?

The days proceed. I strike a match to light the candle in the cave of my heart. It is no easy task, my uneasy truce between the past and future, judgment and acceptance, fear and hope. Fortunately, to begin, there is Carl, who reminds me on occasion that permaculture does not necessitate adopting a donkey and two alpacas whose manure would add to the health of the soil. Or pigs to rout out all the stones that grow far more successfully than potatoes. Or goats to clear the brush and maintain our wildlife

Mom's sculpture "Rising Figure" poolside

186

Rainbow picture of garden and coop area

habitat (though we do rent them). Or sheep for their milk and wool. Or those really cute Runner ducks, and Cayuga ducks, because ducks would eat the slugs that would eat our as-yet-unspawned mushroom beds. And what about quackaponics!? How picturesque to think of the ducks swimming in water that gives nutrients to romaine and kale and radish sprouts. And us, too. A natural web and cycle.

A permaculture ethos does not necessitate any of this, but it holds the possibility. In the meantime, we plan to adopt a seventh generation. Skeetering balls of fluff that cheep relentlessly, calling for their mother. Me. Until another broody hen arrives, I play that role, hovering and clucking. And when they grow old enough to be in the big coop, I imagine the peeps gathering around Uncle Schtude. They will beg him to tell a story and he will. In his sleepy, fluid way, he will relate the adventures of five hens and a cock-a-doodle-do, and their descendants, and the secret that humans have forgotten that it is the chicks' responsibility to teach.

Once upon a time, my mother and I visited my father's mother at the nursing facility that was connected to her assisted living home. We walked away at a fast clip. My mother looked fierce, outraged, because for her, as for most people, life does not mean dozing in a wheelchair in a hallway that smells of antiseptic.

"Kill me," my mother said, "but don't put me into one of those places."

My mother remains in her home, if with limited capacities. A handful of her friends visit. Most of them, though, are busy with life as people do get. And maybe they are afraid to see her, afraid they will catch it—not Parkinson's necessarily, but old age, the passage from youth to senescence. The irony being, they are missing the essence of her.

I, for my part, struggle when we are apart to understand her, and when we are together, too. Clearly, she knows what she wants to say, but the words, and then the thoughts, disappear.

"I'm sorry, Mom," I say, as I did to my father when I neglected to write down his words, and so they, too, were lost. "I don't understand."

"Don't be obtuse!" she replies, clear as day. Why don't I understand her? We both channel my father's frustration of not being able to hear. I look at her, eyes wide, and then? We laugh because it is what it is, and what else to do?

"We'll talk about it later, after you've rested," I say, and reach for a tissue to wipe away the excess drool. She studies me.

"I'm sorry," she says. To distract, I tell her the most recent chicken drama, or look for some happy news in the newspaper. And then, "Did I fail you?"

She asks me the question that I ask myself. Did I fail my parents? Did I fail in my potential? And what a dangerous question: Did she fail me? I look into her eyes. At times, looking at her, I am overwhelmed by the change, heartbroken by who she was and who she has become. But though she isn't who she was, she is who she is, and as she falls into silence, losing words, unable to articulate her thoughts, even to give a superficial under-standing of them, I lean over and kiss her. I say, "I love you."

She sits up. Her face flushes, her eyes brighten, and her arms rise up to wrap around me.

"Oh! Oh! Oh!" she says. "Me, too. I love you." And we hug, both flu-ent, in the end, in the language of love and being.

Ping on my shoulder

Another round of chicken death. Chickadee, CooLots, and Ping are all gone. I miss them, especially Ping, the friendliest chicken, my buddy. Their deaths proved to me that I could not do what I had promised: I might have given them good lives, but not good deaths, because I didn't give up hope that they might live, and so they suffered. What lesson am I to draw from that, when hope is supposed to be a good thing?

Apricot is the last of the old girls. She, alone of our chickens, knew Big Red and his hens, and is as feral as ever, a trait she encourages in our most recently adopted six Icelandic pullets—Colette, Billie Holiday, Rosie and Flopsie, Toey and Copper—a hardy breed, and troublemakers with their derring-do antics. They dodge and flap over the heavier and slower Squeaky, Splotches, and Susie B. Cady, we lost to a weasel.

As for Schtude, I'm not as close to our current flock because he is aggressive in his protection of them. He tests the limits of Carl's and my patience with his sudden attacks that have left us bruised and scarred. At times, I have wanted to rehome him. Instead, I have learned to watch him eyeing me through his floppy feathers, clearly debating another assault. To prevent one, I reach out to pick him up. He trips over himself to get away. Schtude. He is a beautiful klutz of a bird, and he's just being a rooster. And sometimes, the best of times, he hops up next to me where I sit on a hay bale and, together, we admire the girls as they work. He even lets me pet him, and so I hold hope he will, indeed, become the more fluid, thoughtful Uncle Schtude of my imagination. Thus, I balance deep mourning with hope.

Living on the hill that is Darwin's View, especially when the wind is howling, I am fairly convinced that, some day, a storm will destroy the house and coop, ravage our work, perhaps kill us. In the meantime? The chickens

Another rainbow, at sunset

scratch and peck. The sun powers the batteries that power the house on top of the windy hill that we claim to steward. In fact, it stewards us. We try to listen as the sun sets on Mount Monadnock. Every night, it is a different show, the light, the colors, the sounds. Some nights, there is the eerie keening of a pack of coyotes, a wild medley of yips and shrieks and then abrupt silence, a fearsome loneliness and isolation.

Nature echoes here. Her vast vocabulary, like my mother's, has been lost, but the sense memory exists. The pattern shimmers of the web that used to hold us, reflections of the extinct. The haunting beauty and horror of what's been lost, and the world announcing it, is bigger than us as the moon rises, bringing her own brilliance to the night, her procession across the star-speckled sky, an alternative way of being.

However long we are gifted the time here, in the end, nature will take this place and its view back. In fifty years, Carl and I will most likely be dead. We have no children. Only dreams. Only our four hands to create what we will. Our footprint that, with time, will disappear like the trail of a horseshoe crab in the sand on the beaches I grew up near, eroded by waves.

190

The bifrost, spoken of in Norse mythology, is the rainbow bridge between the human world and that of the gods. I think of it as the connection between our limited consciousness and nature's vast awareness. I see it as the bridge between past and future. At Darwin's View, where we see double rainbows and sunspots, we can almost see that bifrost. It has a gleaming edge we call the cutting edge. It is strengthened with blood and tears. We will never get out of Darwin's View the money we have put into it. Fortunately, that is not, and never has been, our intention. Clearly. This is the place where we live. Here we will try to share what we have learned, once we figure out what we've learned. Politics, permaculture, chicken rites and passages, working to return our democracy back to The People, the people back to the land, the land back to its health, health back to the world, the world with its vast beauty. Our home as nature ramps up the ante. In the background, gunshots from the gun club. The wind howls in harmony with the coyotes, the birds, the peepers.

Looking about them at the view, at the place, people ask me "Do you love it?" as if to say, "How could you not?"

I allow for a pause and a heartbeat. Love is a strong word and my attitude toward Darwin's View is balanced by a breathless fear of the elements we face and a dread of the future. Must one love one's home?

If so, why have we humans destroyed the earth?

Why are we doing nothing to fix it?

Whys echo through the mountains and my heart. And we are an adaptable species. Darwin's View is not what I would have expected to be home. It is uncomfortable in my heart's core. But it speaks to the part of me I need to listen to more closely in order to reconnect to the earth and to myself. I can't imagine doing that in any other place. I cannot imagine any other life than this one.

And so. Early on in our Darwinian adventure, I attended an environmental writers' conference in Vermont. While there, I chatted with one of the facilitator/teachers. She and I circuited the topic of home, a topic about which we both wrote. She said her home was near Boston but that she had sold that farm and now lived where she lived, which was not near Boston. She admitted where she lived was not truly home. I didn't understand how she could be so relatively content living with that absence. Years later, I did. I do.

ACKNOWLEDGMENTS

To Mount Monadnock as representative of Mother Earth, for her endless beauty, power, and wind. Her lessons are boundless.

Thanks to Jane Eklund, Ann Hood, Lynn Stegner, and Joanne Wyckoff for their editing and feedback on the book over the years.

To Sarah Bauhan, Henry James, and Mary Ann Faughnan, for their energy, humor, and wit, and willingness to take on my book with all its kerfuffles. They are my publishing family and home.

To my myriad circles of friends and family in Rhode Island and New Hampshire, East Coast to West Coast, past and present. So many have helped me to adapt to and appreciate this life that Carl and I lead. I cannot begin to express the depths of my gratitude for your friendship, love and support.

One circle, in particular, I must name: the incredible people who have supported and cared for my mother. They encompass a world in and of themselves: Denise Ayala, Dawna Milton, Laverne Brown, Sherma Thomas, Kelly Takemura and the rest of the Select Care team; Elissa Kleinhaus, Dr. Alessandro DiRocco and Dr. Christopher Paredes, Theresa Bavero, Amy Lemen, Vashti Greene, Ebony Claxton, Tammy Croteau and her gals, Tara Sabino, and so many others who have come and gone with great caring and respect for, and appreciation of, my mother. Thank you.

And then the caretakers of the cats and chickens: Dr. Mary Coffey of Veterinary House Call Services; Dr. DeVinne and all the technicians and assistants at the Animal Care Clinic; Dr. St. Lifer of Upstream Animal Hospital; and Dr. DeSena and all the technicians of the Marlborough Veterinary Clinic. Your patience with my worries has been as unbounded as is my gratitude for your help.

As ever, my sister, my mother, and Carl.

And, of course, Big Red, his hens, and their descendants, without whom my life would have been far less interesting. Chickens are characters and develop character in those who attempt to steward them. May they and their kindred sentients be treated with the respect and care they deserve as we humans step forth into the new and crazy world we have created.